AF320440

aus der Reihe:

Innovationen mit Mikrowellen und Licht

Forschungsberichte aus dem Ferdinand-Braun-Institut, Leibniz-Institut für Höchstfrequenztechnik

Band 71

Yi-Fan Tsao

Transceiver Technologies for Millimeter-Wave Beam Steering Applications

Herausgeber: Prof. Dr. Günther Tränkle

Ferdinand-Braun-Institut   Tel.   +49.30.6392-2600
Leibniz-Institut   Fax   +49.30.6392-2602
für Höchstfrequenztechnik (FBH)
Gustav-Kirchhoff-Straße 4   E-Mail fbh@fbh-berlin.de
12489 Berlin   Web   www.fbh-berlin.de

Innovations with Microwaves and Light

**Research Reports from the Ferdinand-Braun-Institut,
Leibniz-Institut für Höchstfrequenztechnik**

Preface of the Editors

Research-based ideas, developments, and concepts are the basis of scientific progress and competitiveness, expanding human knowledge and being expressed technologically as inventions. The resulting innovative products and services eventually find their way into public life.

Accordingly, the *"Research Reports from the Ferdinand-Braun-Institut, Leibniz-Institut für Höchstfrequenztechnik"* series compile the institute's latest research and developments. We would like to make our results broadly accessible and to stimulate further discussions, not least to enable as many of our developments as possible to enhance everyday life.

This work deals with design, implementation and test of innovative transceiver modules for future 6G beam steering communication systems at mm-wave frequencies. All module components were designed to achieve high efficiency and drastically reduce complexity. This called for innovative solutions in system and circuit design and finally in the technological system implementation. Thus, the work focused on the development of all rf-circuit components. These included low-noise power amplifiers (LNPA) and single-pole-double-throw (SPDT) mm-wave switches necessary for operating a beam steering system concept based on planar dual exponentially tapered slot antennas (DETSA). The final transceiver module was successfully verified at 38 GHz using an over-the-air test arrangement.

We wish you an informative and inspiring reading

Prof. Dr. Günther Tränkle
Scientific Managing Director

The Ferdinand-Braun-Institut

The Ferdinand-Braun-Institut researches electronic and optical components, modules and systems based on compound semiconductors. These devices are key enablers that address the needs of today's society in fields like communications, energy, health and mobility. Specifically, FBH develops light sources from the visible to the ultra-violet spectral range: high-power diode lasers with excellent beam quality, UV light sources and hybrid laser systems. Applications range from medical technology, high-precision metrology and sensors to optical communications in space and integrated quantum technology. In the field of microwaves, FBH develops high-efficiency multi-functional power amplifiers and millimeter wave frontends targeting energy-efficient mobile communications as well as car safety systems. In addition, compact atmospheric microwave plasma sources that operate with economic low-voltage drivers are fabricated for use in a variety of applications, such as the treatment of skin diseases.

The FBH is a competence center for III-V compound semiconductors and has a strong international reputation. FBH competence covers the full range of capabilities, from design to fabrication to device characterization.

In close cooperation with industry, its research results lead to cutting-edge products. The institute also successfully turns innovative product ideas into spin-off companies. Thus, working in strategic partnerships with industry, FBH assures Germany's technological excellence in microwave and optoelectronic research.

Transceiver Technologies for Millimeter-Wave Beam Steering Applications

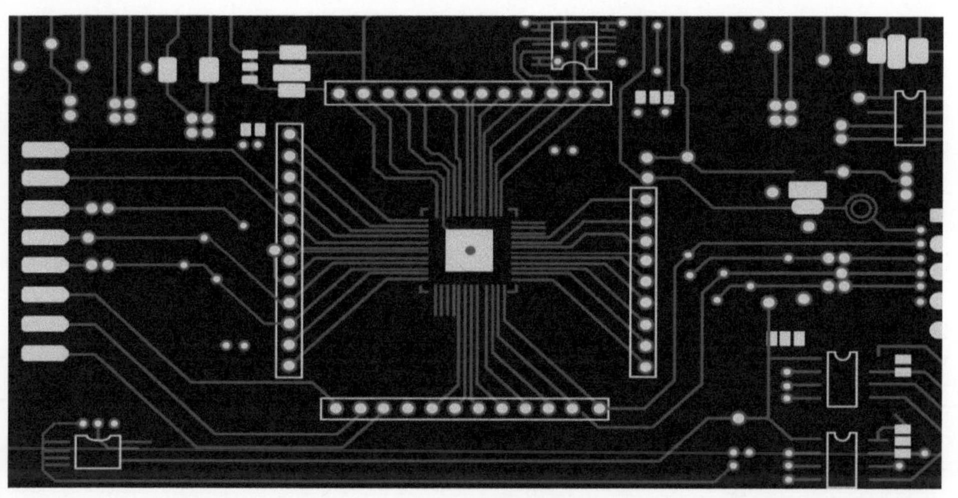

Yi-Fan Tsao - Dissertation

Transceiver Technologies for Millimeter-Wave Beam Steering Applications

vorgelegt von
M.S.
Yi-Fan Tsao
ORCID: 0000-0001-6601-8308

an der Fakultät IV - Elektrotechnik und Informatik
der Technischen Universität Berlin
zur Erlangung des akademischen Grades

Doktor der Ingenieurwissenschaften
- Dr.-Ing. -
genehmigte Dissertation

Promotionsausschuss:

Vorsitzender: Prof. Dr. Günther Tränkle
Gutachter: Prof. Dr.-Ing Wolfgang Heinrich
Gutachter: Dr.-Ing Hans-Joachim Würfl
Gutachter: Prof. Dr. Edward Yi Chang
Gutachter: Prof. Dr. Heng-Tung Hsu

Tag der wissenschaftlichen Aussprache: 18. Juli 2022

Berlin 2022

Bibliografische Information der Deutschen Nationalbibliothek
Die Deutsche Nationalbibliothek verzeichnet diese Publikation in der
Deutschen Nationalbibliografie; detaillierte bibliographische Daten sind im Internet
über http://dnb.d-nb.de abrufbar.
1. Aufl. - Göttingen: Cuvillier, 2022
   Zugl.: Berlin, Technische Universität, Diss., 2022

© CUVILLIER VERLAG, Göttingen 2022
   Nonnenstieg 8, 37075 Göttingen
   Telefon: 0551-54724-0
   Telefax: 0551-54724-21
   www.cuvillier.de

   ISBN 978-3-7369-7702-0
   eISBN 978-3-7369-6702-1

Abstract

During the past years, wireless communication systems have been rapidly advancing to meet the high data-rate requirements of various emerging applications. Since peak data rate and channel capacity are proportional to the operation bandwidth, acquisition of enough bandwidth has become the most critical issue for the next-generation communication systems. As the operating frequency expands into the millimeter-wave bands, research efforts have been devoted to address the issue of high free space path loss. Transceiver design approaches have been widely reported for different applications, such as the user frontend and the base station. However, the existing transceivers were normally demonstrated using CMOS compatible technologies, which delivered a relatively low equivalent isotropically radiated power (EIRP) in a small unit cell. Moreover, the particular device characteristics are limiting the linear region for operation. Therefore, it is the main focus of this dissertation to present and discuss new design method for transceiver having the potential to solve these issues. The designs rely commercially-available GaAs and GaN based device technologies.

To reduce the complexity of the transceiver module for further phased-array scaling, a low-noise power amplifier (LNPA) design approach is designed using 0.15-μm GaN-on-SiC high-electron mobility transistor (HEMT) technology. With the utilization of traded off interstage matching topology between loss and bandwidth, the conversion loss induced by the interstage matching network was effectively reduced. As the GaN-based LNPA delivered a large output power, the RF switches to be integrated for the transceiver must maintain high power handling capability. A stacked-FET configuration was adopted for enhancing the allowed voltage swing at high power operation to improve the power handling of the single-pole-double-throw (SPDT) switch. Also, further improvement on the isolation bandwidth was investigated using theoretical analysis on the intrinsic effect of the passive HEMTs. Such intrinsic effect was mainly contributed by the intrinsic parameters of the devices extracted using MATLAB programming.

With the successful implementation of the RF front-end circuits, transceiver modules were integrated on Rogers RO3010 substrate. The planar dual exponentially tapered slot antenna (DETSA) phased-array system showed a compact size with simple biasing network comparing with the conventional transceiver approach. The presented transceiver module was characterized with an over-the-air test at a distance of 1 m, overcoming the free space path loss of 64 dB. It also shows a high flexibility for further integration with a larger number of array systems, which is very promising for the future 5G communication system.

Zusammenfassung

In den letzten Jahren wurden drahtlose Kommunikationssysteme schnell weiterentwickelt, um die hohen Datenratenanforderungen neu aufkommender Kommunikationssysteme zu erfüllen. Da die Spitzendatenrate und die Kanalkapazität proportional zur Betriebsbandbreite sind, ist der Erwerb einer ausreichenden Bandbreite zum kritischsten Problem für die Kommunikationssysteme der nächsten Generation geworden. Daher dehnensich die Betriebsfrequenzen in die Millimeterwellenbänder aus. Konsequenterweise werden Forschungsanstrengungen unternommen, um das Problem des hohen Verlusts bei der Freiraumausbreitung anzugehen. Für verschiedene Anwendungen, wie zum Beispiel Benutzerterminal und Basisstation wurden daraufhin geeignete Lösungsansätze zum Entwurf von Sendeempfängern veröffentlicht. Die existierenden Transceiver wurden jedoch normalerweise mit CMOS kompatiblen Technologien demonstriert, die eine relativ niedrige equivalent isotropically radiated power (EIRP) in vergleichsweise kleine Basiszellen liefern. Darüber hinaus begrenzen die spezifischen Eigenschaften von mm-Wellen Leistungstransistoren in Verbindung mit der dazugehörigen Schaltungstechnik den linearen Betriebsbereich.

Das Hauptaugenmerk dieser Doktorarbeit zielt daher auf neue Entwurfsverfahren für Transceiver unter Verwendung von kommerziell erhältlichen monoltisch integrierten III-V Schaltungstechnologien (MMIC). Die Arbeit zeigt das Entwurfsverfahren, stellt die Schaltungskonzepte vor und charakterisiert und dieskutiert sie.

Um die Komplexität von Transceiver-Modulen für eine weitere Skalierung von elektroniosch schwenkbaren Antennen (Phased Arrays) zu reduzieren, wurde als Designansatz der sogenanntren rauscharme Leistungsverstärker (LNPA) unter Verwendung einer 0.15-μm GaN-on-SiC HEMT Technologie entwickeltweiterverfolgt. Dieses Konzept vereint den rauscharmen Vorverstärker für den Empfangszweig mit dem Leistungsverstärker für den Sendezweig in einem MMIC. Durch das Design einer zweistufigen Zwischenstufen-Anpassungstopologie mit niedrigen Verlustenwurden Anpassverluste effektiv reduziert. Da der GaN-basierte LNPA eine große Ausgangsleistung liefert, müssen die in den Transceiver zu integrierenden HF-Schalter eine hohe Belastbarkeit aufweisen. In diesem Zusammenhang wurde die Stacked-FET-Konfiguration verwendet, um den zulässigen Spannungshub bei Hochleistungsbetrieb zu erhöhen und um somit das Leistungsschaltvermögen des Sende-Empfangsschalter Schalters Single-Pole-Double-Throw (SPDT) Konfiguration zu erhöhen. Außerdem wurde eine weitere Verbesserung der Isolationsbandbreite unter Verwendung einer theoretischen Analyse des intrinsischen Effekts der passiven HEMTs untersucht.

Mit der erfolgreichen Implementierung der RF-Front-End Schaltungen wurden Transceiver-Module zusammen mit einem DETSA Antennensystem auf einem Rogers RO3010 Substrat integriert. Das planare DETSA Phased-Array-System zeigte eine kompakte Größe mit einem einfachen Vorspannungsnetzwerk im Vergleich zum herkömmlichen Transceiver-Ansatz. Das vorgestellte Transceiver-Modul wurde mit einem Over-the-Air-Test in einer Entfernung von 1 m charakterisiert, wobei der Freiraum-Pfadverlust von 64 dB überwunden wurde. Es zeigt auch eine hohe Flexibilität für die weitere Integration mit einer größeren Anzahl von Array-Systemen, was für das zukünftige 5G-Kommunikationssystem sehr vielversprechend ist.

List of Contents

Acknowledgements

I would like to appreciate for the invaluable opportunity, tremendous support and monumental guidance from my advisor, Dr. Hans-Joachim Würfl. His encouragement and suggestions during my study at Ferdinand Braun Institute have made the time a memorable experience in my life. My Ph.D. study won't be great and wonderful without his effort. He also gave me an opportunity to have research experience in different countries. I'm glad to have him as my advisor.

I would also like to thank my co-advisor, Dr. Heng-Tung Hsu. I can never be at this stage without him. As a teacher and as a friend, his pleasant personality and innovative ideas have made my research experience not only limited in certain categories. Without his advice and assistance, there could not be too much significant progress in my research. I'm glad to have him as my advisor in both study and life. Furthermore, I want to thank Dr. Günther Tränkle for providing the opportunity to join TU Berlin and FBH for my Ph.D research.

The support from the department of GaN Electronics is also of great importance to my PhD study. I gratefully thank for the accompany and help from all the colleagues, Sergey, Mihaela, Matthias, Oliver, Hossein, Oliver and Richard during my stay at FBH. Moreover, the help from microwave department is also acknowledged. I would like to thank Olof and Bernd for the suggestions on circuit designs and the verification of the MMICs.

Finally, I'd like to express my sincerely thanks to my family. Without their support and encouragement, I won't be able to go through these years of study. Also, the great opportunity to meet my wonderful advisor won't exist without the help from my grandmother. These invaluable help have made the dissertation possible.

List of Abbreviations

General

CMOS	Complementary Metal Oxide Semiconductor
EIRP	Equivalent Isotropic Radiated Power
GaAs	Gallium Arsenide
GaN	Gallium Nitride
InP	Indium Phosphide
LNPA	Low-Noise Power Amplifier
HEMT	High Electron Mobility Transistor
SPDT	Single Pole Double Throw
RF	Radio Frequency
DETSA	Dual Exponentially Tapered Slot Antenna
MMIC	Monolithic Microwave Integrated Circuit
LTE	Long Term Evolution
3G	Third Generation
4G	Fourth Generation
5G	Fifth Generation
6G	Sixth Generation
SNR	Signal-to-Noise-Ratio
NR	New Radio
FR2	Frequency Range 2
ETSI	European Telecommunications Standards Institute
MAG	Maximum Available Gain
ETSI	Maximum Stable Gain
IL	Insertion Loss
S-Res	Steering Resolution
TWA	Traveling-wave Antenna
ILA	Integrated Lens Antenna
BF	Beamforming
SBA	Switched Beam Antenna
R-Arrays	Reflec Arrays
AiP	Antenna in Package
VGA	Variable Gain Amplifier
TSMC	Taiwan Semiconductor Manufacturing Co., Ltd.
OTA	Over-the-air
PCB	Printed Circuit Board
QAM	Quadrature Amplitude Modulation
PA	Power Amplifier
LNA	Low Noise Amplifier
MIMO	Multi-Input Multi-Output
SiGe	Silicon Germanium
pHEMT	Pseudomorphic High Electron Mobility Transistor
RL	Return Loss
Iso	Isolation
PAE	Power Added Efficiency

EVM	Error Vector Magnitude
PRBS11	Pseudorandom Binary Sequence 11
SRRC	Square-Root-Raised-Cosine
EM	Electromagnetic
NF	Noise Figure
DPA	Doherty Power Amplifier

Symbols *(e.g. for mathematical equations)*

P_{1dB}	1-dB Compression Power
P_{sat}	Saturated Output Power
TRx	Transceiver
L_{fsl}	Free Space Path Loss
R_{on}	On-State Resistance
C_{off}	Off-State Capacitance
Z_0	Characteristic Impedance
G_m	Transconductance
f_T	Unit-Current Gain Cutoff Frequency
f_{max}	Maximum Oscillation Frequency
I_{DSS}	Zero-Gate-Voltage Drain Current
NF_{min}	Minimum Noise Figure
Γ_{opt}	Optimum Source Impedance for NF_{min}

List of publications

[1] Tsao, Y.-F.; Würfl, J.; Hsu, H.-T. Bandwidth Improvement of MMIC Single-Pole-Double-Throw Passive HEMT Switches with Radial Stubs in Impedance-Transformation Networks. *Electronics* 2020, 9, 270.

[2] H. T. Hsu, T. J. Huang, Y. F. Tsao and J. Würfl, "Bandwidth Improvement of Slot Antenna Arrays Using Substrate Integrated Waveguide (SIW) Through Optimal Design of Radiating Slots," *2019 IEEE Asia Pacific Microwave Conference (APMC)*, Singapore, 2019.

[3] Yi-Fan Tsao, Chien-Ming Tsao, Heng-Tung Hsu and Joachim Würfl, "A High Power Ku- to Ka-Band Single-Pole-Double-Throw Switch with Capacitive Loading for Isolation Improvement," *2020 IEEE MTT-S Latin America Microwave Conference (LAMC 2020)*, 2021, pp. 1-3.

[4] Y. Wang, Y. -F. Tsao, C. -M. Tsao, H. -J. Würfl and H. -T. Hsu, "A Polarization Switchable Antenna Switch Module for Ka-band Application," *2021 IEEE International Symposium on Radio-Frequency Integration Technology (RFIT)*, 2021, pp. 1-3.

[5] Y. -F. Tsao, C. -M. Tsao and H. -T. Hsu, "A New Design Technique for Power Amplifiers Operating Close to Unit-Current-Gain Cutoff Frequency," *2021 IEEE International Symposium on Radio-Frequency Integration Technology (RFIT)*, 2021, pp. 1-3.

[6] Y. -F. Tsao, Y. Wang, C. -M. Tsao, H. -J. Würfl and H. -T. Hsu, "An X- to Ka-band Single-Pole-Double-Throw Switch with Good Power Handling Capability," *2021 IEEE Asia-Pacific Microwave Conference (APMC)*, 2021, pp. 229-231.

1. Introduction

1.1. Motivation

Wireless communication technology showed a fast growth rate during the past decades. Starting from the 1980s, research efforts have been devoted to the development of wireless communication systems. Fig. 1.1 shows the evolution of wireless communication system. While the key performance indices are predicted values, the 6G white paper recently published recommends 140 GHz as the system frequency with a peak data rate of 100 Gbps to 1 Tbps. Applications using wireless communication technology such as the End-to-End architecture shown in Fig. 1.2 were increasing rapidly during the past years. Obviously, the requirement of data rate for each generation of the communication system has increased severely with the thriving of innovative applications. For instance, mobile applications in fourth-generation Long-Term-Evolution (4G LTE) networks and smart city targeted for the fifth-generation (5G) communication system. These data-hungry applications are the driving forces for the tremendous evolution of wireless communication systems which proceeds in ever shorter evolution cycles.

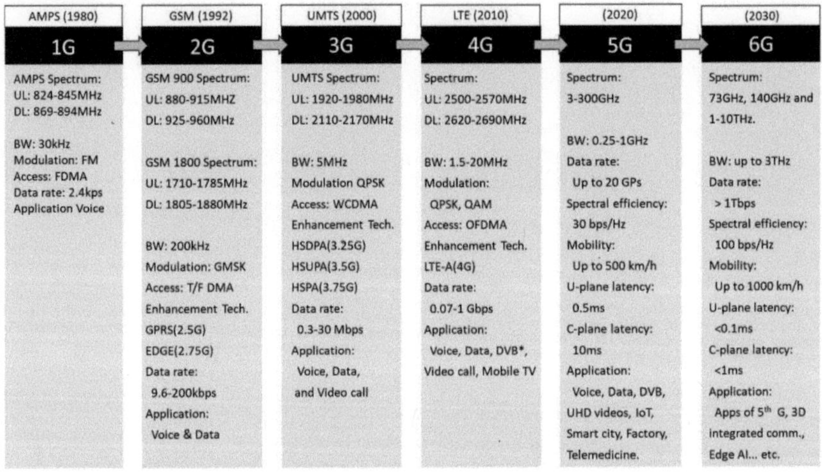

Figure 1.1. Evolution of wireless communication system [1].

Fig. 1.3 shows the timeline of wireless communication systems. A duration of 15 years from standardizing the specification to deployment has been taken for the third-generation (3G) communication system; as for 4G, the duration was 12 years. For the ongoing 5G communication system, an estimation of 8 years from standardizing to full deployment has been made. Communication system technology beyond 5G is expected to have even shorter duration as shown in the figure. While targeting for a higher data transmission rate, the available spectrum for the implementation of communication systems seems to be a critical issue. This is mainly attributed to a crowded spectrum at sub-6 GHz frequencies with a variety of commercial applications. To seek for higher spectral efficiency, moving up to millimeter-wave frequencies is considered as a potential solution.

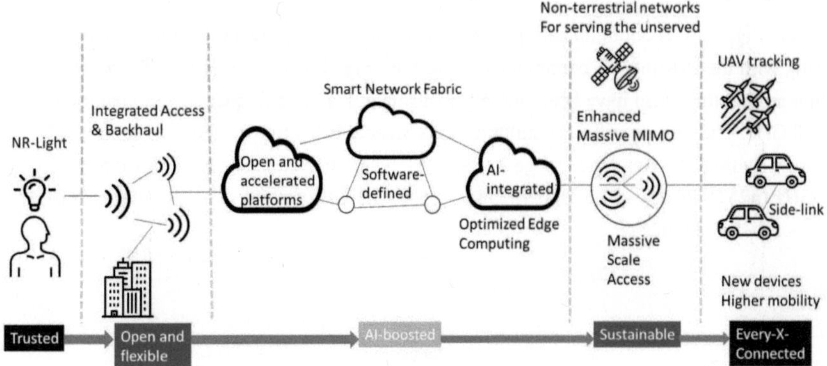

Figure 1.2. End-to-End (E2E) architecture [2].

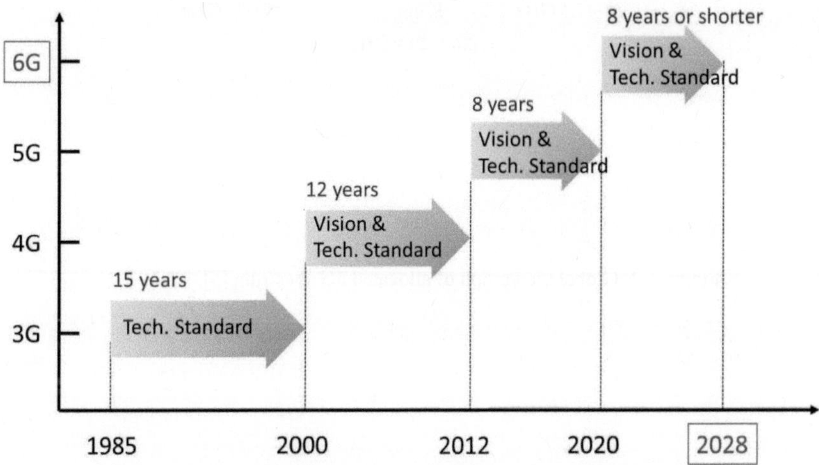

Figure 1.3. Timeline of wireless communication systems [3].

Referring to the Shannon-Hartley theorem [4], a wider operating bandwidth and high signal-to-noise ratio (SNR) are required for the realization of high capacity 5G wireless network. The peak data rate and capacity of a radio channel can be expressed as:

$$\text{Peak data rate of a radio channel} = BW \cdot n \cdot M \tag{1.1}$$

$$\text{Capacity of a radio channel} = BW \cdot n \cdot \log_2 (1 + SNR) \tag{1.2}$$

Note that BW stands for bandwidth, n stands for the number of spatially separated paths and M represents the number of bits. The two equations highlight the importance of channel bandwidth for enhancing performance of communication system. Fig. 1.4 shows the frequency allocation for 5G communication systems over the world.

As shown in Fig. 1.1, 5G communication system serves a higher bandwidth and lower latency comparing with the existing 4G communication system. Fig. 1.3 shows that the 5G spectrum allocation around the world have been majorly focusing at Ka-band frequencies in vicinity of 28 and 38 GHz. The European Telecommunications Standards Institute (ETSI) has announced the 5G New Radio (NR) frequency range 2 (FR2) [6], where five channels are defined as 26.5 – 29.5 GHz (n257), 24.25 – 27.5 GHz (n258), 39.5 – 43.5 GHz (n259), 37 – 40 GHz (n260) and 27.5 – 28.35 GHz (n261).

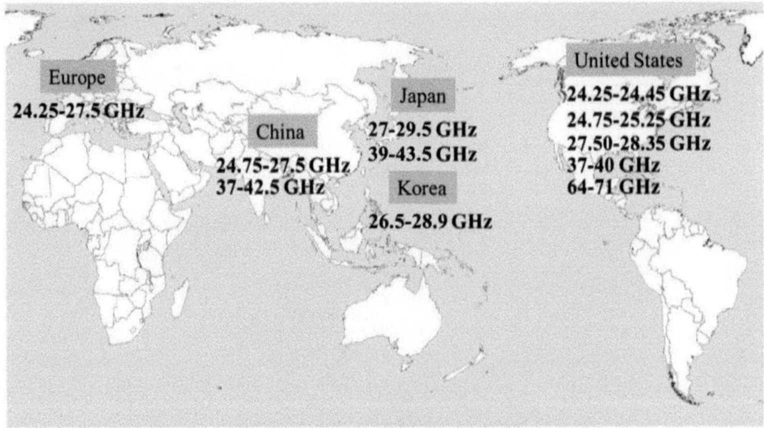

Figure 1.4. Global viewgraph of allocated 5G spectrum [5].

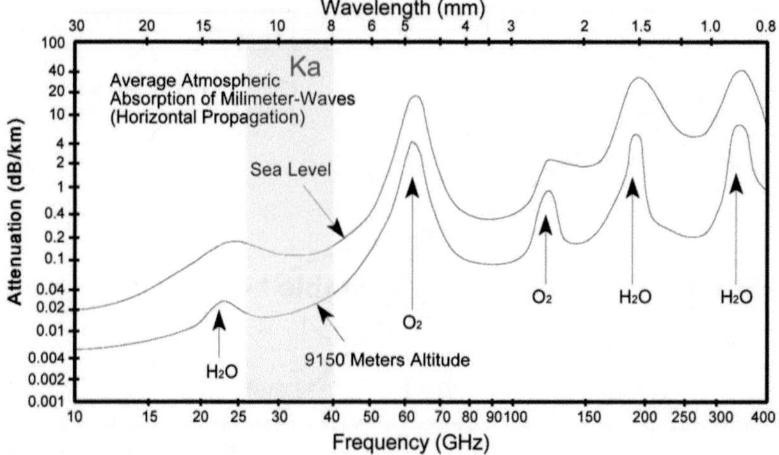

Figure 1.5. Average atmospheric attenuation at millimeter-wave frequencies [10].

For millimeter-wave applications, it can be expected that high attenuations such as atmospheric loss [7], dielectric loss in substrates [8] and metallic loss in conductors [9] severely influence amplifier efficiencies. Such drawbacks have led to the difficulties in the deployment of the basic infrastructure. Fig. 1.5 shows the average atmospheric absorption at millimeter-wave frequencies from 10 GHz up to 400 GHz, where Ka-band frequencies are marked in light blue.

To overcome the high attenuation at millimeter-wave frequencies, enhancement in the overall performance of the communication system in terms of gain and output power are considered as a critical issue. However, the severe degradation of maximum available device gain (MAG) as shown in Fig. 1.6 at high operating frequencies makes it difficult to meet system specifications. To further increase the MAG at higher operating frequency, scaling down the gate length of devices in conjunction with a systematic reduction of parasitic capacitive, inductive and ohmic losses in the devices are considered as straightforward methods. Fig. 1.7 shows the status of popular device technologies implemented for millimeter-wave and submillimeter-wave circuits. Although the maximum operating frequency has been pushed to a very high level, the shorter gate length is expected to limit the breakdown voltage of the devices. The output power can be expressed as:

$$P_{out} = \frac{\left(I_{D,\max} - I_{D,\min}\right) \times \left(V_{D,\max} - V_{D,\min}\right)}{8} \tag{1.3}$$

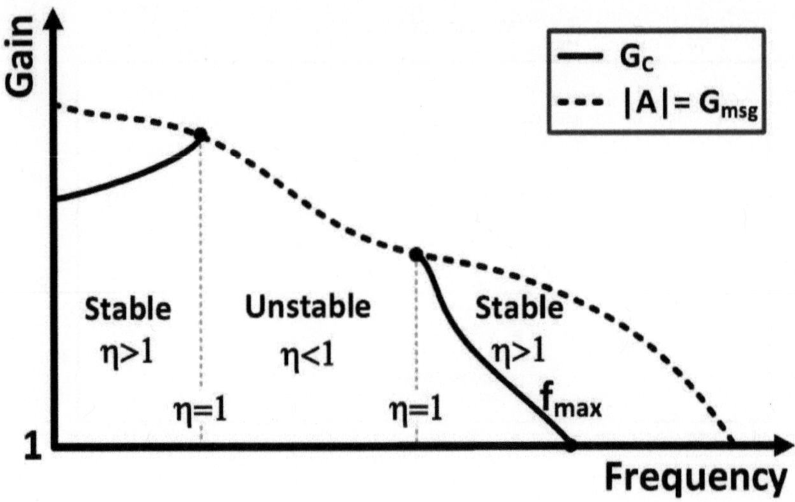

Figure 1.6. Maximum available gain and maximum stable gain as an interval of frequency [11].

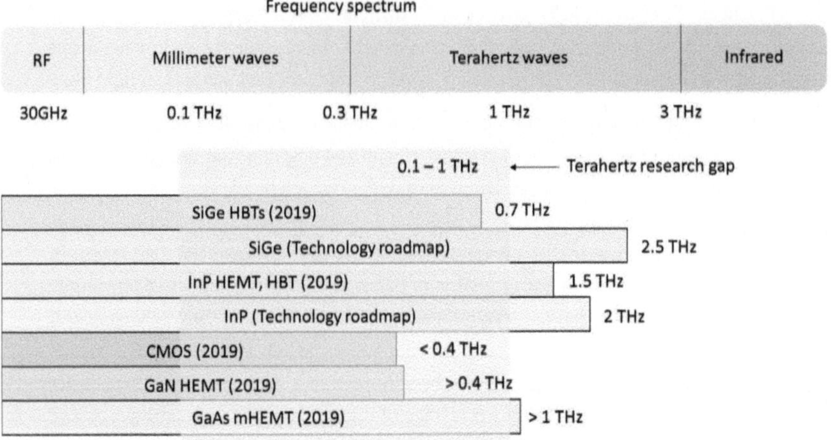

Figure 1.7. Maximum operating frequency of different device technologies [12].

Therefore, the breakdown voltage of the device may limit the maximum drain voltage allowed for the proper operation. According to eq.1.3 this also limits the maximum output power of the device. To seek for higher output power density at higher frequencies, compound semiconductor device technologies with relatively high breakdown voltage characteristics are considered as ideal candidates for applications at millimeter-wave frequencies and beyond. Fig. 1.8 plots the saturated output power (P_{sat}) of amplifiers as a function of frequency in log scale using different device technologies. Focusing at the frequency range between 10 to 100 GHz, device technologies with higher breakdown voltage such as gallium arsenide (GaAs) and gallium nitride (GaN) are showing outstanding performance in terms of output power at millimeter-wave frequencies comparing with CMOS technologies. As for indium phosphide (InP), it is showing a nice performance for applications beyond 300 GHz. Regarding the trend for GaN-based amplifiers, the P_{sat} of the reported power amplifiers at frequency above W-band drops severely. Comparing with mature III-V technologies including GaAs and InP, GaN technologies are still under development for operating frequencies beyond W-band. For recent publications, GaN-based power amplifiers beyond W-band frequencies were majorly focusing on the gain improvement instead of output power. This was mainly due to the relatively low frequency response characteristics comparing with GaAs and InP technologies, leading to a lower MAG level. The output power level was not a big issue for GaN-based amplifiers as it was stated that the high breakdown voltage allowed a higher voltage supply at the drain node, which resulted in a greater output power density. Therefore, it can be inferred that the slope of the trend can be smoothed soon in the future with the improvement of frequency response and device structure optimization.

Although each of the device technology has its shortage while pushing the limits of maximum operating frequency, systematical design methods have been widely reported for circuitries at high frequencies [14]-[17].

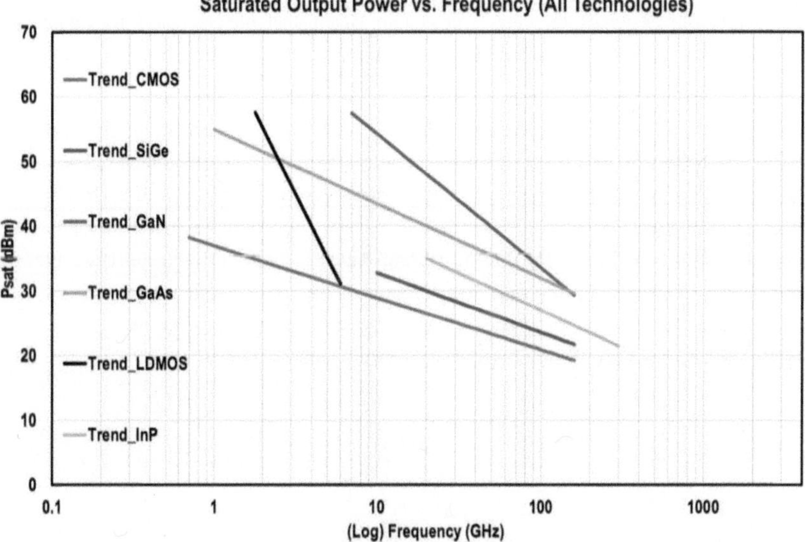

Figure 1.8. Saturated output power of amplifiers as a function of frequency using different device technologies [13].

To push boost up the deployment of 5G communication system, design technologies taking care of the specific requirements dictated by the semiconductor technologies applied have to be taken into account. Hence, transceivers with high gain, high power for overcoming the atmospheric attenuation are necessary to support the operation at commercial bands regulated for 5G NR FR2.

1.2. Requirements for beam steering mm-wave transceivers

For millimeter-wave applications, the propagation characteristics affecting the transmitting and receiving of transceivers have been discussed in the previous section. Based on the discussion, the first issue to solve will be the free space path loss (L_{fsl}), which is expressed as [18]:

$$L_{fsl} = 32.44 + 20\log f + 20\log R - G_{Tx} - G_{Rx} \tag{1.3}$$

Note that f represents frequency in GHz, R is the distance between antennas in m, G_{Tx} and G_{Rx} are the overall transmitter/receiver antenna gain including feeding loss. For long distance transmission, the free path loss is considered as an extreme challenge for evaluating the link budget for the transceiver system. To overcome the issue, antennas with high gain and high directivity are promising candidates to solve the issue. However, the atmospheric attenuation is not the only application concern for transceiver at millimeter-wave frequencies. Obstacles such as walls and other objects are known to induce high penetrating losses at millimeter-wave frequencies [19]. Therefore, plenty of issues regarding the induced losses must be carefully considered prior to the design of the transceiver system. Beam steerable antennas have caught researchers' eyes as the ability of reconfiguring radiation patterns to maintain proper signal transmission in the network.

Beam steering antennas were first implemented at low frequencies for systems requiring high directivity beams to prevent unwanted signals. Take a receiver with omnidirectional radiation pattern as an example, the signals from each device may cause interference with each other while receiving the signal of interest. To properly receive the desired signal without the interference of unwanted signals, the beam steering antennas are able to reconfigure the propagation beam in such a way that the transmission of the desired signal approaches an optimum. Such arrangements improve system performance as the interference has been reduced. Fig. 1.9 shows the diagram for illustration of beam steering with multiple devices.

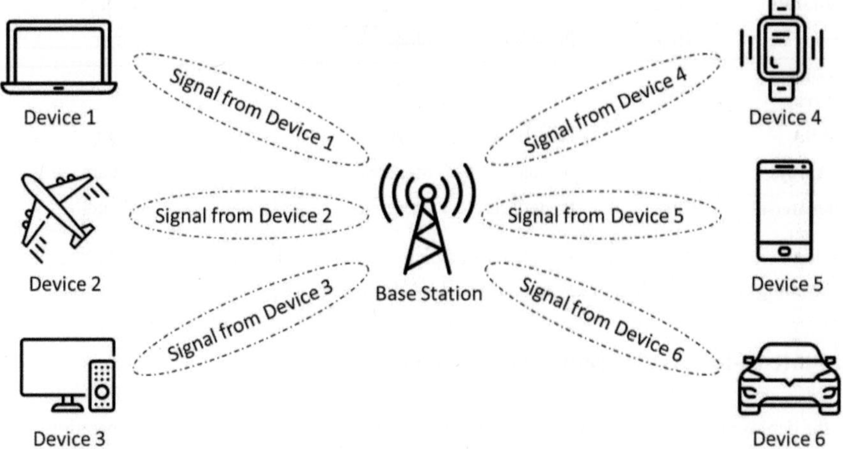

Figure 1.9. Illustration of beam steering approach.

Beam steering technique has some important properties to maintain for operation:

1. When an object is presenting the transmitting path, an additional insertion loss is induced and degrades the transmission of power delivered to the system.

2. The steering range of a beam steering antenna system is another concern for design. It depends on particular design and implementation of the individual antennas and of the phase shifter network. For example, a horn antenna using non-uniform flexible meta-surface allows for a steering range of the main beam of 80° as reported in [20].

3. The steering resolution (S-Res) of the beam steering antenna system can be set as continuous, predefined or finite. The setting will affect the increment of the steering range, which mainly depends on the needed steering angle for the system.

4. The sensitivity of the beam steering system will be determined by the steering speed. However, this depends on the environment for operation.

Table 1.1 presents an overview of the beam steering techniques [21]-[30]. A variety of research utilizing different beam steering techniques have been published in conferences or journals, showing a nice performance at millimeter-wave frequencies.

Table 1.1. Comparison table for common beam steering techniques.

Technique	IL	S-Res	Complexity	Size	Cost
TWA	None	Continuous	Low	Small	Low
ILAs	Low	Predefined	Low	Medium	Low
Parasitic	Low	Predefined	Low	Depending	Low
Mechanical	None	Continuous	Low	Large	Low

Digital BF	High	Fine	High	N/A	High
Analog BF	High	Predefined	Moderate	Medium	High
Reflect Array	Medium	Predefined	Moderate	Large	High
SBA	Medium	Predefined	Low	Large	High
R-Arrays	Low	Fine	Moderate	Medium	Medium
Metamaterial	High	Predefined	Moderate	Medium	Medium

Traveling wave antenna (TWA); Integrated lens antenna (ILA); Parasitic steering; Beamforming (BF); Switched beam antennas (SBA); Reflect arrays (R-Arrays)

1.3. State-of-the-Art in transceiver technology

With the deployment of 5G communication systems, applications are expanding their operating frequency into millimeter-wave region. For high speed data links, phased-array transceivers composed of multiple antenna arrays and RF front-ends are adopted to achieve high equivalent isotropic radiated power (EIRP). Such arrangement with phased-arrays enables multiple applications such as high speed link connecting base station interacting with mobile devices [19], [31]. The 5G commercial spectrum has launched sub-6 GHz systems at the early stage. The combination with already existing 4G communication systems initially reduced complexity. To push the operating frequency up to Ka-band frequencies and beyond, it is necessary for the antenna system to support a large number of controllable beams using cost-efficient arrangements suitable for realistic deployment. In this connection, Global Foundries developed a circuit block of 32 TRx module including RF phase shifting, switched TRx and simultaneous H/V polarization architectures based on their 8HP 130-nm SiGe BiCMOS process. Fig. 1.10 [32] shows the block diagram of this particular module. For the switched TRx, antenna switches controlling the signal path, low-noise amplifier improving Rx sensitivity and power amplifier improving the EIRP of Tx with nice linearity are designed and integrated in a single chip. As for the RF phase shifting architecture, the available phase control range enables beamforming operation, nice phase shifting resolution is required for improving side-lobe suppression to minimize the phase error induced while beam steering. A variable gain amplifier (VGA) was designed to expand the control range of RF gain, which enables a high level side-lobe rejection and minimize the complexity of system calibration.

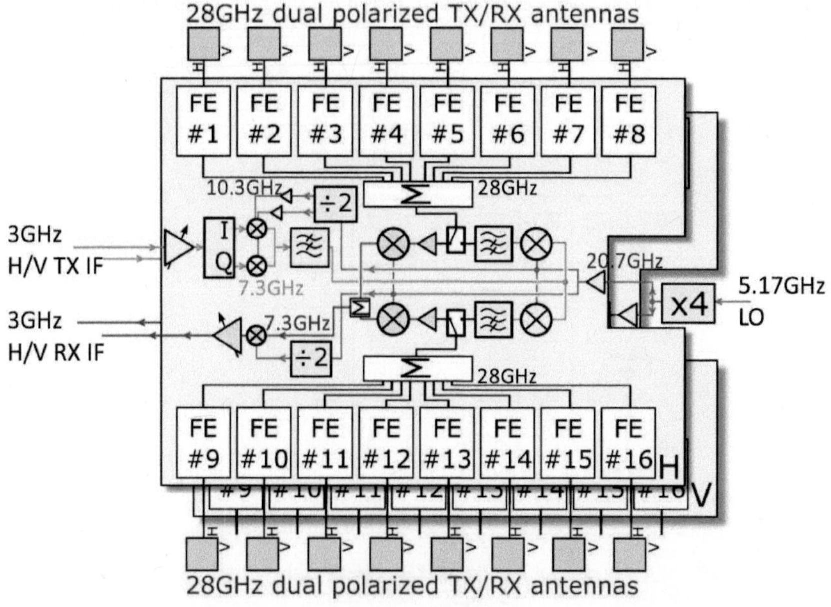

Figure 1.10. Schematic for the building block of TRx module [32].

For the single chip solution, characterization of the transceiver must go through three stages including on-wafer measurements, packaged IC for over the air tests and connectorized package to be wired with a millimeter-wave interface to process further characterizations. The antenna-in-package (AiP) approach has been widely reported in CMOS and SiGe technologies for millimeter-wave applications [32]-[35].

As for antenna on printed-circuit-board (PCB) approaches [36]-[39], it often required a large number of chips for further integration. However, there are several disadvantages for the antenna on PCB approach. For instance, large propagation loss degrading the transmitted power while scaling up the number of arrays. In [39], the 2×2 TRx beamformer chips with symmetric design using Wilkinson power divider/combiner was proposed for overcoming the disadvantage of antenna on PCB approaches. Fig. 1.11 shows the building block of the $N \times N$ phased-array with 2×2 TRx beamformer on PCB. To improve the transmitted power and noise figure of the system, the routing distance between the chip and antenna feeds was minimized for reduction of the conversion loss. The scalable characteristic by using symmetry Wilkinson networks can modify the dimension of the module based on the system specifications. With a large number of elements to be implemented in the phased-array system, heat dissipation induced by the chips are considered as a critical issue as this may have led to degradation of the overall performance. To overcome such issue, apertures have designed on the PCB to spread out the heat without disturbing antenna efficiency. Silver paint with a thermal conductivity of 9.1 W/m·K and low sheet resistance of 0.08 $\Omega/\square \cdot \mu m$ was applied to connect the front and backside of the PCB. Such arrangement helped to maintain the available transmitted power for the phased-array system as the apertures effectively spread out the heat dissipating from the chipsets.

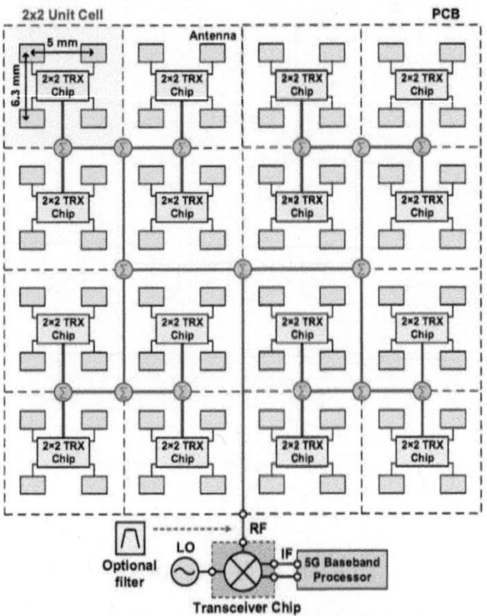

Figure 1.11. Schematic of 64-element phased-array with 2 × 2 beamformer [39].

The TRx chip was fabricated using TowerJazz SBC18H3 0.18-μm SiGe BiCMOS process. However, in order to enhance the performance of the phased-array system, the RF front-end and the transceiver chip were designed separately. While targeting high output power and low noise figure, the RF beamformer was designed using a SiGe process. On the other hand, for the maintenance of baseband performance, CMOS technology was applied for the transceiver chip. As shown in Fig. 1.11, a RF filter was set optional for integration in the system. This was mainly attributed to the level of out-of-band signals, which may lead to the LO radiation in the IF architectures. The high flexibility for scaling and low cost for integration are considered as the main advantage for the antenna on PCB approach. However, a disadvantage in terms of the dielectric and metallic loss with the PCB substrate may require additional gain on the amplifiers to achieve the desired specification. An over the air test of 300 m has been made using 1.6 Gbps 16-QAM modulated signal with the highly scalable transceiver with 2 × 2 beamformer, where 45 dBm EIRP was achieved at a P_{sat}. These two approaches are suitable for beam forming concepts even in the handsets of end-users. However, the EIRP is still limited as, in contrast to GaN and GaAs, CMOS and SiGe technology-based amplifiers are showing a lower output power. Fig. 1.12 shows the large-signal performance with peak PAE and P_{sat} of reported power amplifiers using different device technologies at 20-50 GHz. The reported results have proven that GaAs and GaN are feasible for high EIRP approaches.

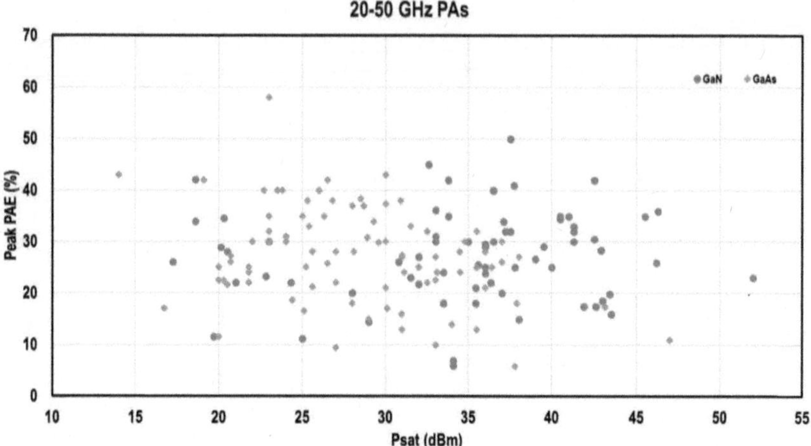

Figure 1.12. Performance of power amplifiers at 20-50 GHz with different device technologies [13].

For base station applications, phased-array system with very high EIRP is necessary for proper functionality. Therefore, separate ICs are required to be integrated on a scalable PCB module as shown in Fig. 1.13. For high power and low-noise amplifiers, GaAs and GaN device technologies are often considered as competitive candidates. However, as mentioned previously, the large dissipated heat from the amplifiers during operation may degrade the overall system. To alleviate the heat issue, a larger size of the PCB is needed with more apertures to spread out the heat. Such arrangement will lead to a bulky size. Despite the disadvantages, the high flexibility is feasible for further extension of the transceiver module for enhancing the transmitting power. To reduce the size of PCB for spreading out the heat induced by active circuits, materials with high thermal conductivity are considered as a solution for the substrate of the carrier, such as AlN (321 W/m·K).

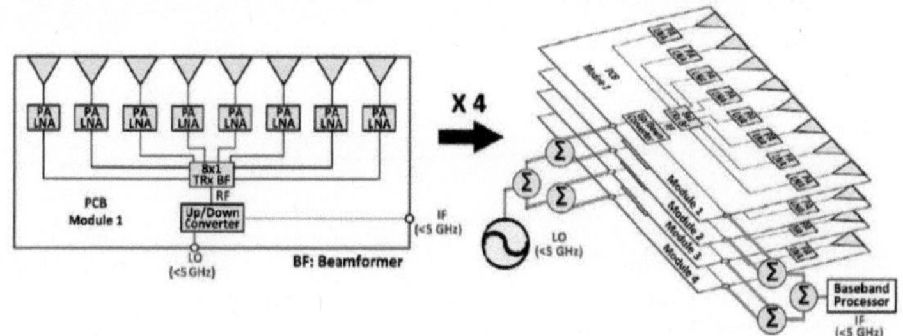

Figure 1.13. Large scalable transceiver with separate-ICs [40].

A phased-array system composed of eight-element Tx and four-element Rx with endfire antennas using separated ICs was presented in [40]. In this phased-array system, the differential I/Q up-converter and beamformer ICs were designed and fabricated using 65-nm CMOS technology provided by Taiwan Semiconductor Manufacturer Company (TSMC). While targeting for high signal-to-noise ratio and EIRP for long distance transmission, 0.15-μm GaAs pHEMT technology were applied for power and low-noise amplifiers. Finally, 16-element Rx and 32-element Tx brick array antenna modules were assembled by wire-bonding techniques and by using a low-cost stacking method: Tx-to-Rx data rates of 0.9 Gbps using 512-QAM modulation scheme were reported for a 20 m over-the-air measurement. Moreover, a beam steering angle of 60° was achieved at azimuth.

1.4. Outline of dissertation

In this dissertation, the design, the implementation and the characterization of transceivers targeting for millimeter-wave beam steering applications are reported.

Chapter 2 will go through the components to be integrated in the RF front-end transceiver system. A review of the RF switch and low-noise amplifier design approaches will be discussed. Next, design procedure and analysis of two single-pole-double-throw (SPDT) switches and a low-noise power amplifier will be reported. Furthermore, the experimental verification based on 0.15-μm GaAs pHEMT and 0.15-μm GaN HEMT technology will be presented.

Chapter 3 discusses antenna configurations for millimeter-wave applications. A design and technological implementation with the butler matrix for beam steering application will be presented. Module integrations with the designed MMICs reported in chapter 2 will be included for further characterization.

Chapter 4 highlights the potential applications for the presented transceivers in this dissertation, illustrating that the design methods can meet the requirements of next generation communication system.

2. Design and Implementation of Transceivers

This chapter provides an overview on the targeted transceiver system and explains design methodologies for suitable RF front-end circuits.

2.1. Transceiver system overview

The transceiver system is designed for transmitting and receiving signals. Fig. 2.1 shows the block diagram of a conventional RF front-end T/R module. It consists of a pair of single-pole-double-throw (SPDT) switches, a power amplifier (PA), a low-noise amplifier (LNA) and an antenna. In modern 5G communication system, phased-array T/R modules are required to be implemented in both user equipment and access point. Dimensions of the T/R will be limited by the targeted applications. As higher EIRP level is required for millimeter-wave applications, amplifiers using compound semiconductor device technologies are considered as the key enables. However, this requires a large-scale integration as separated ICs are needed for baseband operation. Moreover, a large-scale array integration based on conventional T/R modules has to address the following challenges: (1) complicated DC biasing network, (2) bulky in overall size and (3) difficult for array extension. Therefore, a compact T/R module is required for 5G and beyond communication systems. Fig. 2.2 (b) shows the block diagram of the proposed compact T/R module.

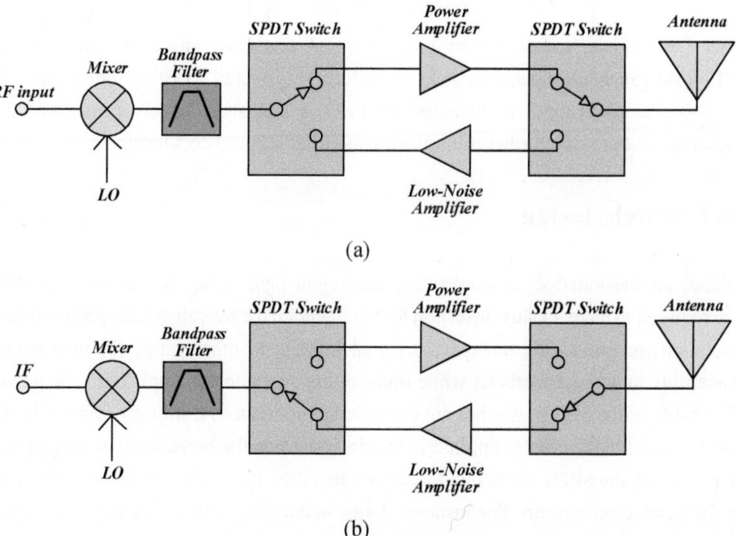

Figure 2.1: Block diagram of a conventional T/R module operating in (a) transmitting mode and (b) receiving mode.

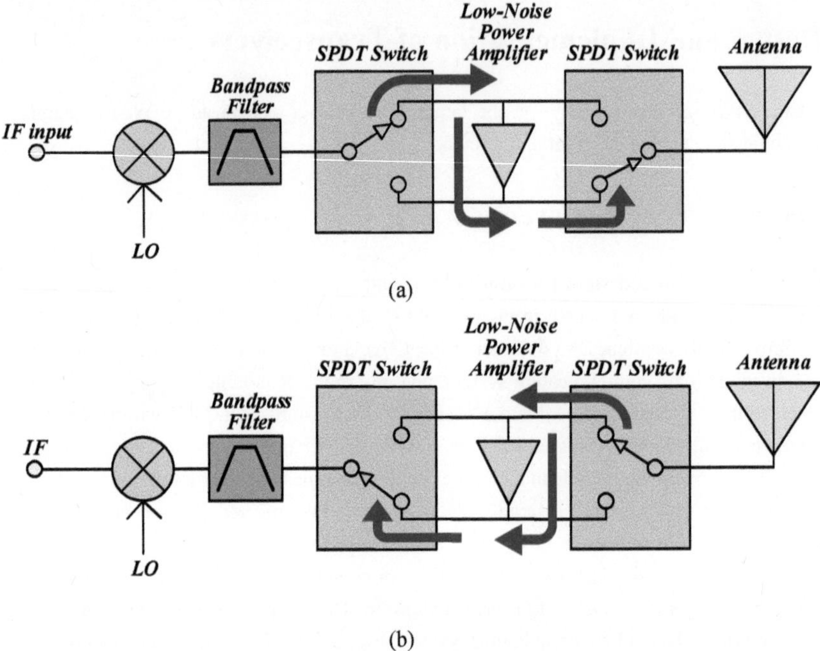

Figure 2.2: Block diagram of a compact T/R module operating in (a) transmitting mode and (b) receiving mode.

The compact T/R module is designed by combining PA and LNA into a low-noise power amplifier (LNPA). Such arrangements help to reduce the overall circuitry dimension. In addition, as massive multi-input multi-output (MIMO) systems are widely considered the proposed solution will provide further minimized systems at reduced cost levels.

2.2. SPDT switch design

SPDT switches are responsible for controlling the signal path in a T/R module. Ideal RF switches are required to have: (1) low insertion loss to maintain the overall system performance, (2) high isolation for preventing unexpected signal leakage to other paths, (3) nice power handling capability to avoid distortions while transmitting signal to the amplifiers. The power handling capability of the SPDT switches are considered important as it strongly referred to the breakdown voltage of the transistor applied in the design. Once the breakdown of the applied transistors is reached, the SPDT switch will lose the capability of distributing signal, leading to damage in the transceiver system. For instance, large signal flows into a low-noise amplifier and overdrives it to the saturation level, leading to the damage of the overall transceiver system.

Section 2.2.1 presents a review of design approaches for millimeter-wave RF switches. Next, design and implementation of wideband, high isolation and high-power handling Ka-band

SPDT switches using 0.15-µm GaAs pHEMT technology provided by WIN Semiconductor will be discussed in section 2.2.2. The experimental verification of fabricated SPDTs using on-wafer probing will be explained in section 2.2.3.

2.2.1. Switch configurations

Different switch configurations are principally possible. Which one of thje approaches dis optimum depends on system specification and design. Basic configurations are:

- Series-type switch [41]-[42],

- Shunt-type switch [43]-[44]

- Series-shunt type switch [45]-[46]

They are commonly used for RF switch designs, the schematic of the three types of switches are shown in Fig. 2.3. As each of them shows independent pros and cons in terms of the performance, design methodologies need to be developed to cope with the specific advantages and disadvantages of each basic configuration.

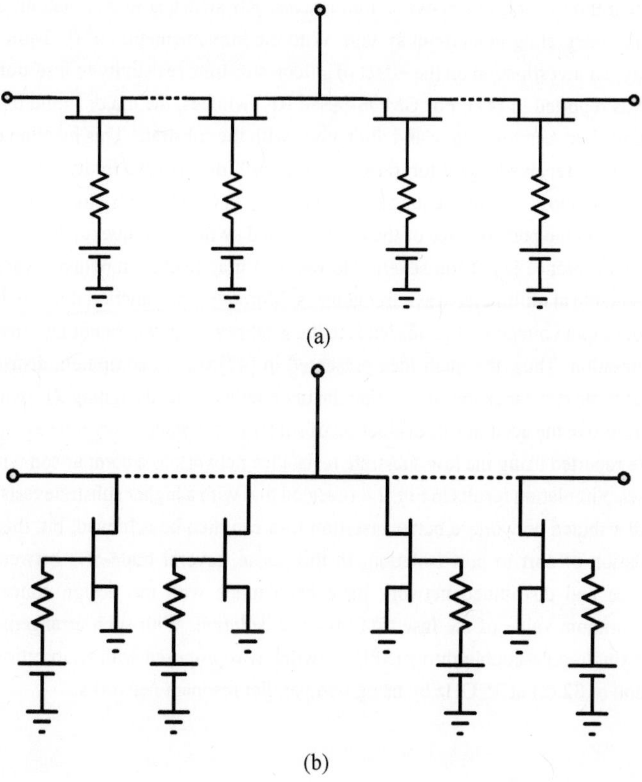

(a)

(b)

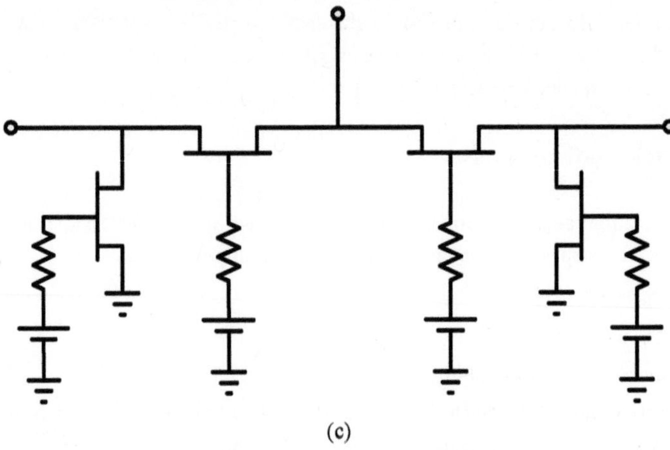

(c)

Figure 2.3: Schematic of SPDT switches using basic configurations (a) series-type, (b) shunt-type and (c) series-shunt type.

Targeting for integration at millimeter-wave frequencies, RF switches must maintain a low insertion loss while integrating in a circuit system. With the implementation of 0.13-μm SiGe CMOS technology, an investigation on the effect of silicon substrate resistivity to insertion loss of a RF switch was reported in [47]. For CMOS-based RF switches, the major limitation will be the junction diode between the source and drain node with the substrate. This junction diode may increase the loss of received signal for transmission at millimeter-wave frequencies, which leads to a large insertion loss. As for the junction capacitance in CMOS transistors, it normally has less contribution on the performance of the CMOS-based switch at frequency lower than 6 GHz, as the phase difference is not too severe. However, it may lead to impedance variation which is not neglectable at millimeter-wave frequencies. Moreover, the junction diode will also reduce the level of output voltage swing, this leads to the weakness of power handling capability at high power operation. Thus, the main idea presented in [47] was to compute a distributed network of the substrate resistance for minimizing the uncertainty while designing RF switches, which helped to improve the accuracy of evaluation of substrate resistance. Several cases of the RF switches were reported using the low substrate resistance network as a comparison with the conventional cases. Simulation results in Fig. 2.4 revealed that with a higher substrate resistance induced by the distributed network, a better insertion loss can then be achieved, but this may lead to a degradation of port to port isolation. In this sense, several trade-offs between the substrate resistance and distributed network have been made with the design concern of approaching an optimum value of the insertion loss and isolation. With such arrangement, a series-shunt type single-pole-double-throw (SPDT) switch was proposed with an insertion loss of 2.2 dB, isolation of 32 dB at 35 GHz by using two parallel resonant networks.

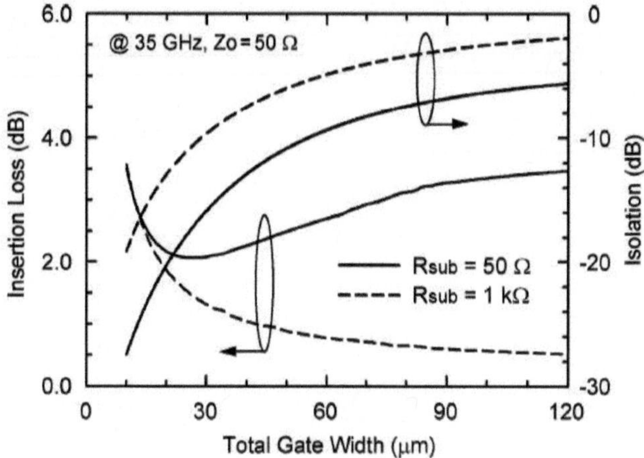

Figure 2.4: Simulated insertion loss and isolation of an nMOS switch as a function of gate width with different substrate resistance at 35 GHz, [47].

2.2.2. Design rules for optimizing SPDT performance

2.2.2.1. Reduction of insertion loss

In order to reduce the insertion loss of the SPDT switch, a design approach using bandpass filter integrated concept was proposed in [48]. Such arrangement also showed an advantage in terms of an overall size reduction of switch MMIC chip area. Starting with the basic topology of the shunt-type switch, the major cost in chip area will be the quarter-wavelength transmission line applied for distributing the signal received at the common input port. Therefore, a π-type resonant structure was proposed using a short serial transmission line with two shunt capacitances to equivalent the quarter-wavelength transmission line. Next, the design procedure started with synthesizing the bandpass filter, which will be applied for the benchmark comparing with the size-reduced bandpass filter. Three different size reduction equations on the size of the bandpass filter have been reported in the paper. The design flow of the bandpass filter with the design equations were reported in [48] as shown in Fig. 2.5. Fig. 2.5(a) shows the original bandpass filter without size reduction, each of the quarter-wavelength transmission line had a synthesized characteristic impedance at the targeted design frequency. With the derivation using the first size-reduction method, the equivalent circuit was plotted as shown in Fig. 2.5(b), where the reduction began with the series quarter-wavelength transmission lines. Next, the second size reduction method were applied to reduce the dimension of the shunt transmission lines be the derived capacitance from the first design method. Finally, the synthesized bandpass filter was applied for matching network with the transistors for final size-reduced filter-integrated SPDT switch. The main advantage of the reported SPDT switches using bandpass filter integrated approach was to minimize the insertion loss by matching the common input and the output impedance of the switch branches to the optimum impedance.

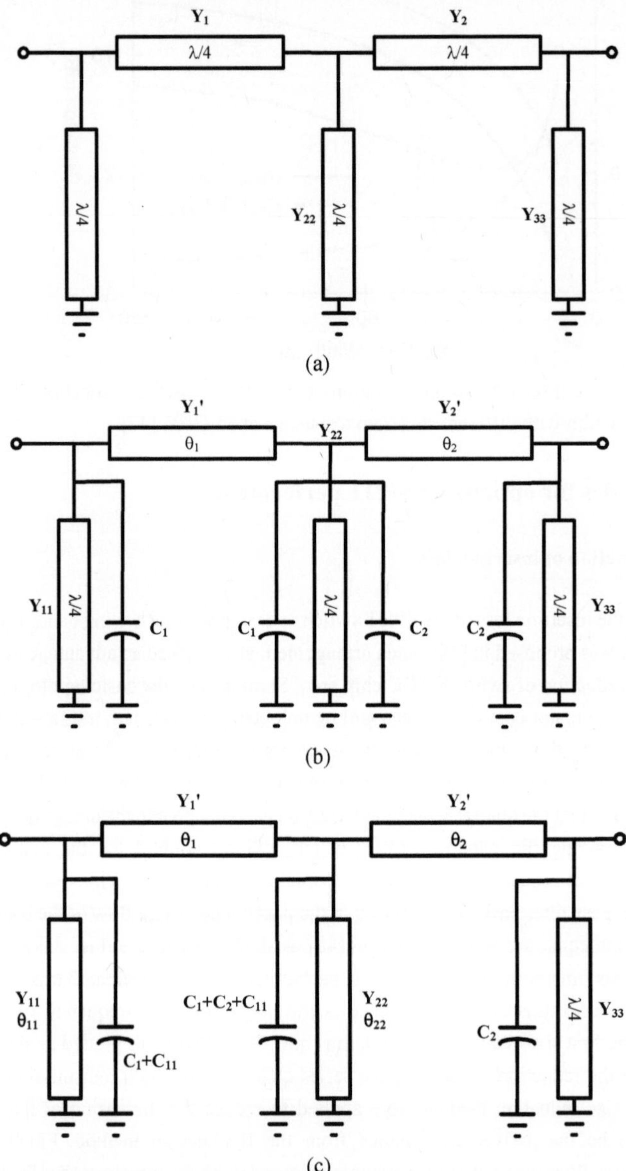

Figure 2.5: Schematic of (a) original bandpass filter, (b) equivalent circuit by using first size-reduction method, (c) equivalent circuit by using second size-reduction method.

The proposed filter-integrated SPDT switch showed an insertion loss less than 1.5 dB, isolation better than 22 dB across the operating frequency of 40-60 GHz. A minimum insertion loss of 1 dB was observed at 40 GHz, exhibiting an outstanding performance comparing with the reported SPDT switches for millimeter-wave applications.

2.2.2.2.   Optimizing switch isolation properties

Insertion loss is not the only design concern for a RF switch, the isolation between output ports are also critical for design implementation for switches with multiple throws. Design approaches for isolation improvement were also widely reported in the previous works. A 60 GHz high isolation SPDT using shunt pHEMT resonator was proposed in [49] to reduce the on-state resistance of the devices. Conventionally, a simplified model of the on-state device for shunt-type switch application can be considered as a resistance shown in Fig. 2.6(a); the off-state device can then be considered as a capacitance shown in Fig. 2.6(b). The main ideal of the work was to replace the conventional open circuit stub shunt at the output ports by a shunt pHEMT resonator, the simplified shunt pHEMT resonator was shown as Fig. 2.7. For operation at the off-state, the low impedance of the shunt device made the shunt FET with transmission line as an open circuit stub. As for on-state operation, the high impedance made the resonating pHEMT as a lossy circuit improving the isolation of the SPDT switch. Such arrangement can help to improve the isolation by 5 dB, as a 0.2-dB reduction for the insertion loss was obtained by utilizing such configuration. The proposed SPDT switch using shunt pHEMT resonator achieved an insertion loss of 1.4 dB and isolation of 45 dB at 60 GHz.

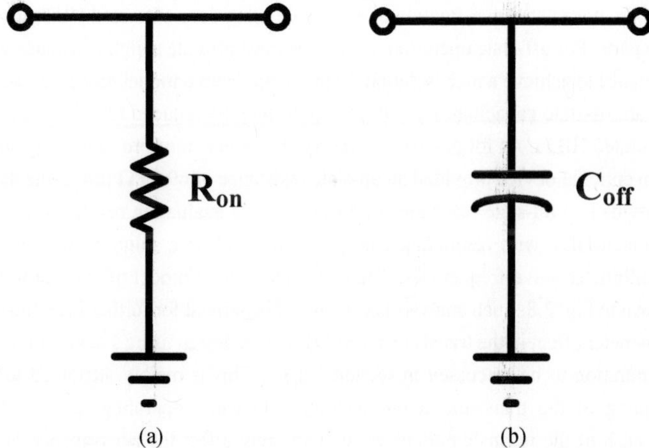

Figure 2.6: Simplified device model of (a) on-state device, (b) off-state device.

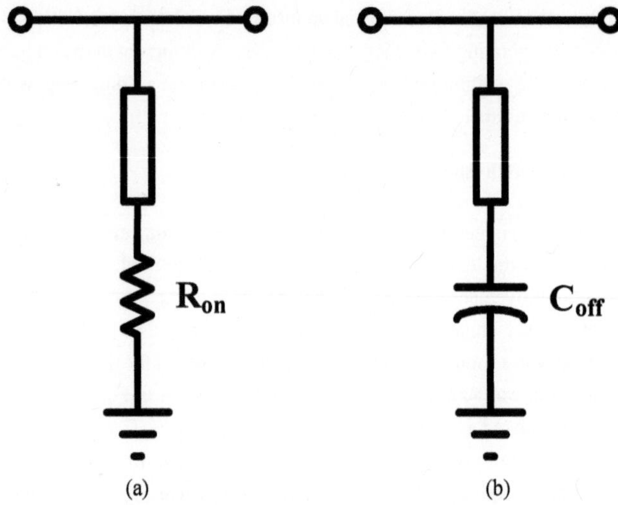

Figure 2.7: Simplified device model of (a) on-state switch, (b) off-state switch using resonant type configuration with a quarter-wavelength transmission in series.

The isolation performance of the RF switches can also be improved by other design approaches. A SPDT switch using deep-saturated SiGe heterojunction bipolar transistors (HBTs) was reported in [50] to enhance the isolation performance. Conventionally, as the HBT device was driven into deep-saturated region, a slow transient effect may be induced leading to the shortage of charging at the base node. The investigation of advantage using deep-saturated HBTs can be separated into two parts. For off-state operation, the device must provide a high resistance with a minimum capacitance to achieve a nice isolation. In this sense, conventional diode-connected HBT switch with an off-state capacitance 3.3 fF/μm could then be reduced to 1.55 fF/μm by using the deep-saturated HBTs. As for on-state operation, the device must provide a very small resistance, the conventional device provided an on-state resistance of 69.75 Ω·μm as the deep-saturated HBT provided an on-state resistance of 54 Ω·μm. The evaluation results evidenced that the deep-saturated HBT with resonating configuration could be a solution for isolation improvement at millimeter-wave frequencies. The equivalent circuit model of each condition was plotted as shown in Fig. 2.8. Such analysis is considered beneficial for further investigation of the intrinsic parameter effect in the transistor for SPDT switch design using stacked-FET and series-shunt configuration to be discussed in section 2.2.3.2. This is mainly attributed to the small-signal modeling of the transistor at on- and off-state while operating as a realistic switching device, each of the intrinsic parameters will strongly affect the performance of the switch as the operating frequency increases. Thus, not only the small-signal characteristics can be determined, but also the large-signal operation can be taken care of. With the implementation in 0.13-μm SiGe HBT, the SPDT switch was measured to have an insertion loss less than 3 dB and isolation better than 23.5 dB from 96 to 163 GHz.

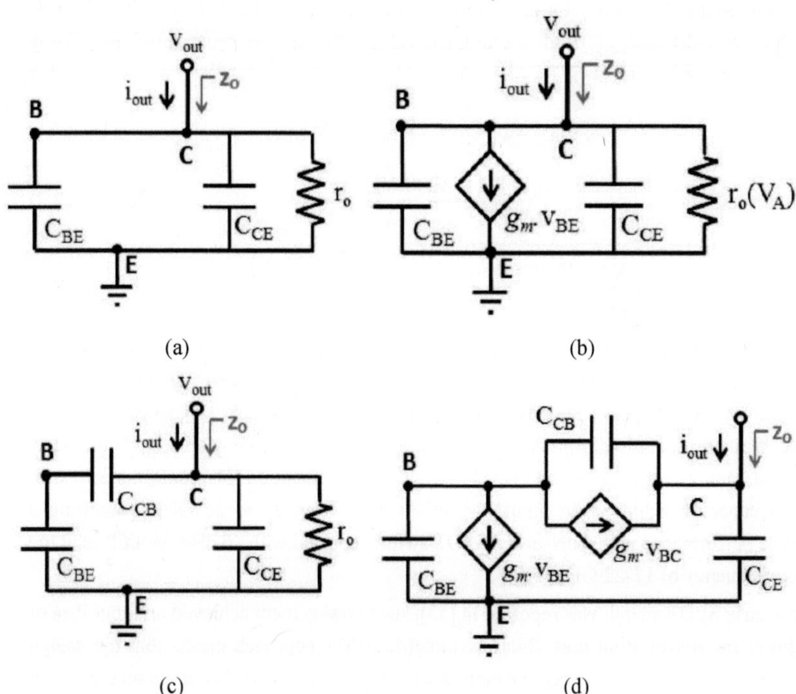

Figure 2.8: Small-signal equivalent device model of (a) diode-connected device at off-state, (b) diode-connected device at on-state, (c) deep-saturated device at off-state, (d) deep-saturated device at on-state [50].

2.2.2.3. Power handling capability

As stated previously at the beginning of this chapter, RF switches are required to have a nice power handling capability to avoid distortions while transmitting signal to the amplifiers. Design methodologies for improving the power handling capability of the RF switches at millimeter-wave frequencies have been proposed during the past years. The stacked-FET configuration was commonly reported for CMOS-based amplifiers for power improvement as it could enlarge the tolerance of output voltage swing at high input power level. In general, with N transistors in a stack, the overall transistor cell can improve its tolerance by a factor of N with a maximum supply voltage defined near to the drain-to-source breakdown voltage specified by the device technology. The stacked-FET configuration also works for the high power handling RF switch design. In [51], the stacked-FET configuration was adopted for enhancing the maximum available voltage swing at the output node of the device to improve the overall power handling capability of the SPDT switch. For commercial 0.15-μm GaAs pHEMT technologies, the breakdown voltage was normally specified as 12 V. To avoid the drain-to-gate breakdown at the common-gate (CG) transistor at high input power level, the supplied operating bias was normally set at 1-dB back-off from the optimal value. The derived equation of the biasing voltage was expressed as (2.1)-(2.3):

$$\left|V_{bias} - V_{Bk}\right| \leq \frac{V_{RF}}{4} \leq \left|V_{Pinchoff} - V_{bias}\right| \tag{2.1}$$

$$V_{Bias\_opt} = \frac{V_{BV} + V_{Pinchoff}}{2} \tag{2.2}$$

$$P_{max} = \frac{V_{RF}^2}{2Z_0} = \frac{8\left[\min\left(\left|V_{Bias} - V_{BV}\right|, \left|V_{Pinchoff} - V_{Bias}\right|\right)\right]^2}{Z_0} \tag{2.3}$$

Based on the given equations, the maximum power handling capability of the switch can be easily identified to be close to the drain-to-source breakdown of the device technology. Once the designer wants to improve the power handling capability, it is necessary to overcome the limitation by using specific design approach. Implemented in the commercial 0.15-μm GaAs pHEMT technology, the proposed SPDT switch using stacked-FET configuration with resonating capacitor achieved an input P_{1dB} of 36 dBm from 22 to 26 GHz, exhibiting a comparable performance with GaN-based SPDT switch reported with 36 dBm input P_{1dB} at the operating frequency of 17-22 GHz [52].

An asymmetric SPDT switch was reported in [53], such arrangement achieved an input P_{1dB} of 31.8 dBm at the transmitting port. Such asymmetric SPDT approach means that the design configuration of the transmit and receive paths are different from each other based on the design purpose, the schematic is shown in Fig. 2.9. The operation of the proposed asymmetric SPDT switch could be separated into the transmitting mode and receiving mode. The simplified equivalent circuit of the transmitting mode and receiving mode were given as Fig. 2.9. While operating at the transmitting mode, the biasing voltage of the two devices were set at on-state, making the simplified model of the devices as small resistors. Such arrangement made the receiver side as a LC tank circuit to achieve high impedance, preventing the leakage of signal to the receiver side in transmitting mode. As for receiving mode, the supplied voltage for both devices were set at off-state to make the simplified device model as a large resistor with capacitance. The operating mode for receiver was similar to the transmitting mode, as the transmitter side could be equivalent as a LC tank circuit achieving high impedance by resonance to prevent signal leakage to the transmitter side. The difference comparing with the transmitting mode operation was that the off-state devices might have limited the power handling capability of the SPDT switch [54]. Although the power handling capability of the SPDT switch was different at different operating modes, it was considered reasonable as the receiver normally required a lower power handling capability comparing with the transmitter. With the implementation in 65-nm CMOS technology, the design exhibited a comparable power handling capability with the III-V technology based designs. On the other hand, the low insertion loss has made this design approach promising for integration with the RF front-end module.

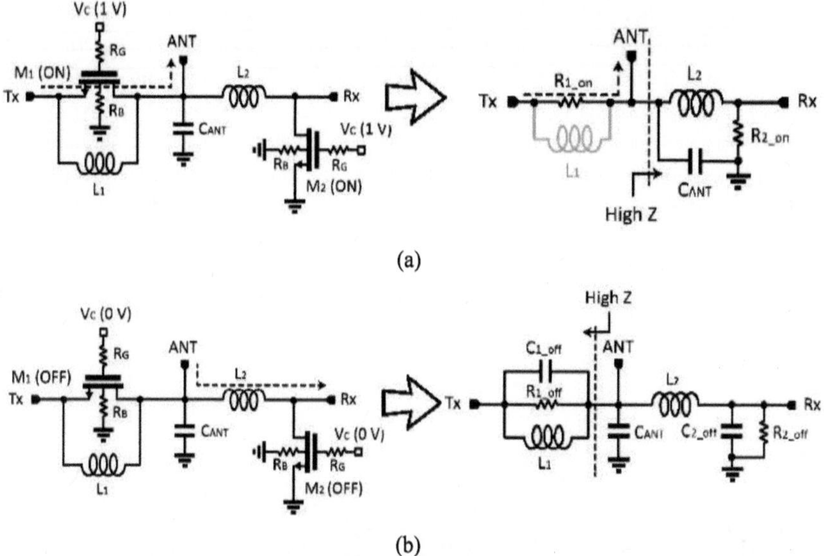

(a)

(b)

Figure 2.9: Simplified equivalent circuit of the asymmetric SPDT switch at (a) transmitting mode and (b) receiving mode [53].

2.2.3. Analysis and Simulations

In this section, two different design approaches of the SPDT switch will be presented. The first part will be a SPDT switch utilizing the impedance transform concept with radial stub (see section 2.2.3.1) for operating bandwidth expansion. The design of a series-shunt type SPDT switch using stacked-FET configuration as explained in section 2.2.3.2 for enhancement of power handling capability will then be discussed.

2.2.3.1. SPDT Switch using Impedance Transformation with Radial Stubs

In this section, a design approach of shunt-type SPDT switch using impedance transformation concept will be discussed. Based on the reported design in [55], such an arrangement was designed to convert the off-state impedance to an ideal short-circuit and on-state impedance to open-circuit for isolation improvement. The main idea was to replace the resonating inductor as reported in [49] by an impedance transform network composed of two fractions of series transmission lines and an open-circuit stub. The resonating structure was reported using a single piece of transmission line to eliminate the parasitic capacitance at the output of the shunt transistors. Therefore, only a narrow frequency range of parasitic capacitance can be compensated, leading to a shortage of operating bandwidth. The impedance transform network was given as shown in Fig. 2.10. The derived impedance transformation equations were also given in [55]. The reported simulation results revealed that the impedance transformation approach could help to improve the performance of isolation and also maintained the insertion loss of the SPDT switch with a proper matching circuit. To achieve an optimum high isolation, the impedance of the off-state transistor at point D must be converted to short condition, which shall be a low resistive impedance (1.4 ohm reported in [55]). On the other hand, the optimum low insertion loss condition shall present the impedance of the on-state transistor at point D converted to open condition. This should be highly resistive impedance (for example 543 ohm reported in [55]). The large on/off impedance ratio between the transmit and receive path of the SPDT switch will help to improve the level of isolation.

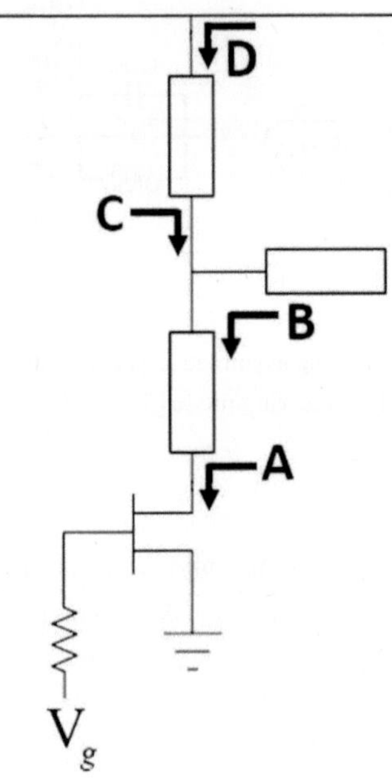

Figure 2.10: Schematic of an FET in series with the impedance-transformation network.

Trade-off between insertion loss and isolation must be made while designing the switch utilizing such design approach. The major reason leading to such issue results in the on-state resistance and the off-state capacitance, while these intrinsic components could affect the on/off state impedance for a device. To evaluate the impedance achieving the optimum insertion loss and isolation for the SPDT switch, a load-pull simulation was made for determining the optimum impedance. The setup for load-pull simulation is plotted as shown in Fig. 2.11. Fig. 2.12 plots the simulated impedance contour for insertion loss and isolation, where V_{g1} and V_{g2} were set to be 0 V and -2 V. For the simulation, the defined areas on the Smith Chart for short-circuit were in the range of [75, 1000] for real part and [-180, 180] for the imaginary part; as for open-circuit, the range was defined for real part as [1, 10] and [-180, 180] for the imaginary part, respectively.

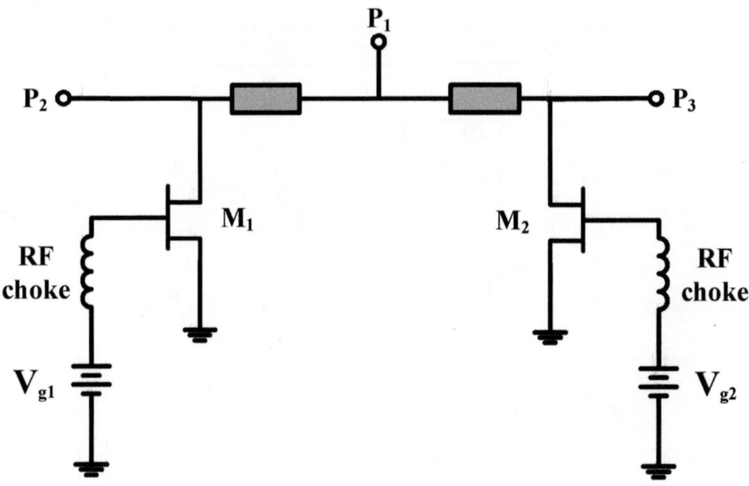

Figure 2.11: Schematic of proposed SPDT switch using impedance transform method.

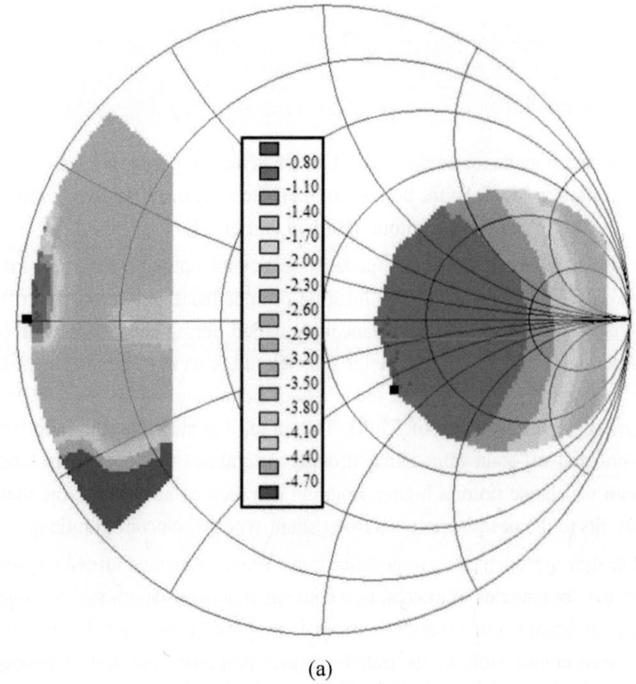

(a)

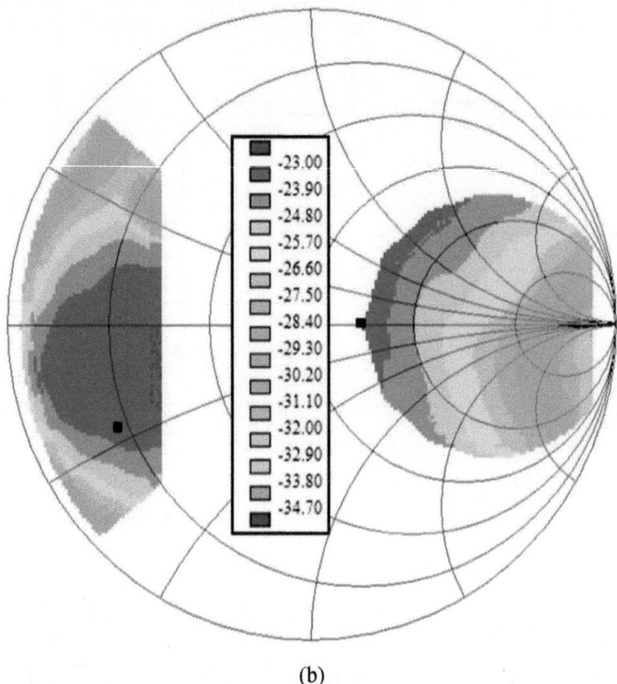

(b)

Figure 2.12: Simulated impedance contour based on the generic topology shown in Figure 2.11 at point D specified in Fig. 2.10 for both shunt transistors (a) Insertion loss contour and (b) Isolation contour. (Both in unit of dB)

From Fig. 2.12, we can observe that the impedance achieving optimum insertion loss and isolation differ from each other. Note that the location of impedance for the on- and off-state cases plotted on the Smith Chart were simultaneously varied during simulation. The lowest insertion loss is observed when the impedance for short-circuit is set at 1 ohm and open-circuit at 75 ohm, but such combination would have led to a poor isolation of 24 dB. This differs a lot from the maximum allowable isolation of 35 dB. Therefore, it is clear that the ratio between real parts of the on- and off-state impedance strongly determines the level of isolation. In general, with a high resistance ratio, a higher isolation can then be achieved. Note that such characteristics only fits to the design approach using shunt-type switch configurations.

The conventional design approach using impedance transformation method turned out not to be practical for mm-wave frequencies of interest as a conventional open-circuit stub was applied. Therefore, a new impedance transformation network is proposed with a radial stub replacing the conventional open-circuit stub as its transformation properties are less dependent on frequency. The proposed impedance matching network is shown in Fig. 2.13. Fig. 2.14 compares the impedance variations of open-circuit stub and radial stub configuration in a frequency range between 25 to 55 GHz. Obviously, the frequency sensitivity of the radial stub is relatively lower comparing with the open-circuit stub.

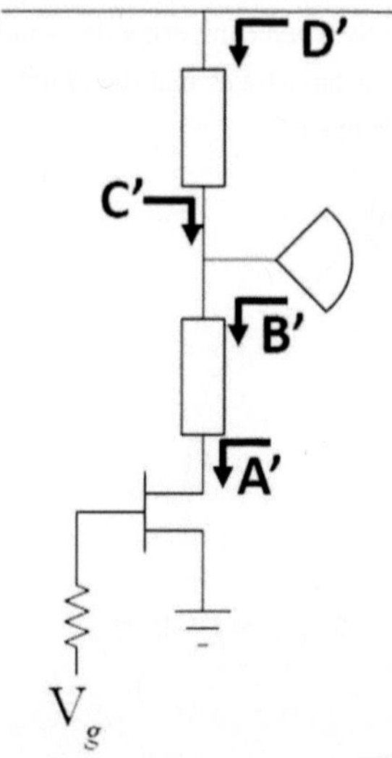

Figure 2.13: Schematic of proposed impedance transform network using radial stub.

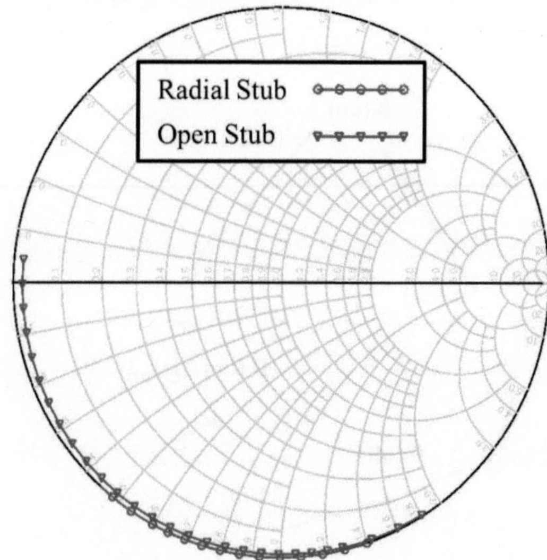

Figure 2.14: Comparison of impedance variation between an open-stub and radial-stub in the frequency range between 25 and 55 GHz.

An investigation of the radial stub in the impedance transform network is also made, Fig. 2.15 plots the trajectories of the conventional and radial stub case on the Smith Chart, where dashed lines represent the impedance of off-state and solid lines for the impedance of on-state. Note that the simulated frequency was set from 25 to 55 GHz and the center frequency was set at 38 GHz. With the comparison per simulated trajectories, a lower angular span and less impedance variations are observed from Fig. 2-15. Especially when comparing the simulations taken from the positions C and D in on-state. This clearly demonstrates that the radial stub truly expands transformation bandwidth and thus operating frequency.

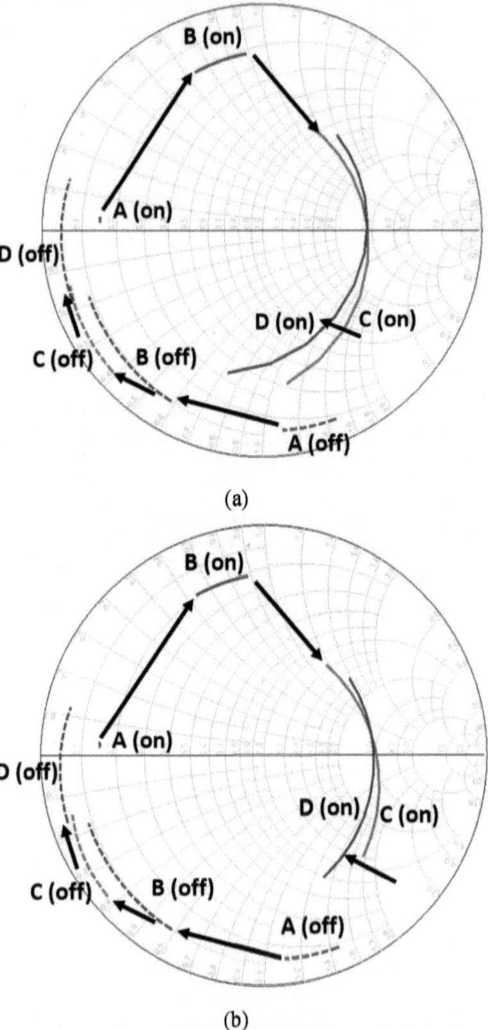

Figure 2.15: Trajectories of the network based on different matching topologies from 35 to 45 GHz with (a) conventional open-stub, (b) proposed radial-stub.

A further investigation of the design parameters for the radial stub will then be performed in comparison with the conventional open-circuit stub. For fairness of comparison, the operating frequency is normalized to 38 GHz, and the simulation frequency is set from 35 to 45 GHz. Based on the simulated results, design parameters of the radial stub can be synthesized with the desired level of specified insertion loss and isolation. For the single stage SPDT design, an insertion loss of less than 1 dB and isolation better than 25 dB was achieved with the synthesized radial stub with a radius of 157 μm and θ of 129°. The impedance trajectories of the effect in terms of radius and angular span are plotted on the Smith Chart as shown in Fig. 2.17. The simulation results reveal that the proposed network can effectively transform the on- and off-state impedance to the optimum impedance for achieving desired insertion loss and isolation.

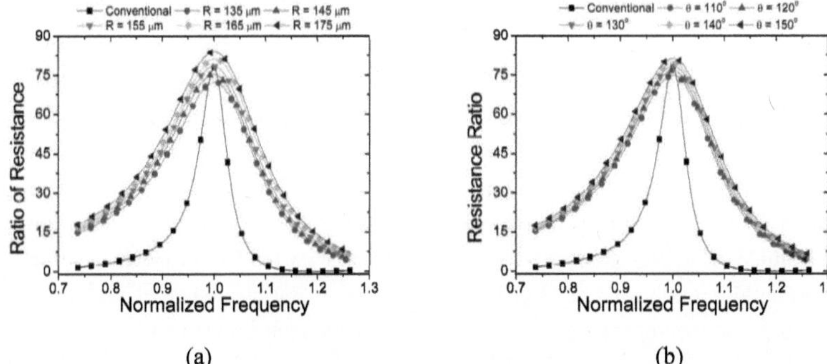

Figure 2.16: Resistance ratio as function of normalized frequency for the cases of (a) fixed θ of 130° with varying R; and (b) fixed R of 155 μm varying θ. The same ratio for the conventional topology using open stub is also included in the plots.

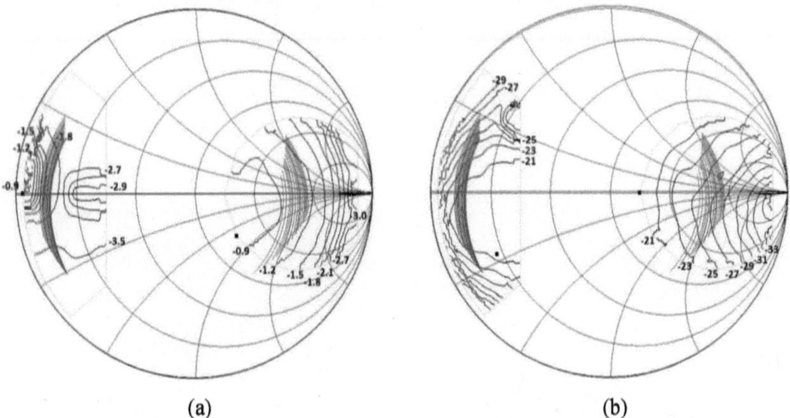

Figure 2.17: Impedance trajectories investigating the effect of design parameters of the radial stub with corresponding contours of the (a) insertion loss and (b) isolation from 35 to 45 GHz.

2.2.3.2.   SPDT Switch Utilizing Stacked-FET and Series-Shunt Configuration

Another design approach using a series-shunt switch configuration will be discussed in this section. Based on the schematic and simplified device model shown in Fig. 2.3(c) and Fig. 2.6, an investigation of the on-state resistance (R_{on}) and off-state capacitance (C_{off}) is performed. The equation of insertion loss (S_{21}) and isolation (S_{23}) are derived as:

$$S_{21} = \frac{4Z_0 R_{on}}{\left(2Z_0 + R_{on}\right) + \omega^2 Z_0^4 C_{off}^2}$$

(2.4)

$$S_{23} = \frac{4\omega^2 Z_0^2 C_{off}^2 R_{on}^2}{\left(Z_0 + R_{on}\right)^2 + \omega^2 Z_0^2 C_{off}^2 \left(2Z_0 + 3R_{on}\right)^2}$$

(2.5)

Note that port 1 is the common input, where port 2 is set as the thru path and port 3 on the isolated path. Fig. 2.18 shows the effect of R_{on} and C_{off} on magnitude of S_{21} and S_{23} at 38 GHz. An optimum combination with low R_{on} and low C_{off} will achieve low insertion loss and high isolation with the series-shunt configuration. Fig. 2.19 plots the magnitude of S_{21} and S_{23} with a fixed C_{off} of 30 fF. The simulation results reveal that R_{on} must be small enough to achieve a maximum magnitude in S_{21} and minimum magnitude of S_{23}. With the fixed capacitance, the minimum insertion loss is achieved at 20 GHz. As for isolation, the magnitude of S_{23} shows inconspicuous variation with operating frequency. Similar simulation results with a fixed R_{on} of 2 ohm are plotted in Fig. 2.20. In Fig. 2.20 (a), the magnitude of S_{21} shows an implicit variation with operating frequency. As for isolation, the magnitude of S_{23} shows a large variation while changing the capacitance at different operating frequencies. Therefore, for wideband SPDT switch using series-shunt configuration, proper selection of the device size will strongly affect the characteristics. To maintain reasonable performance at Ka-band frequencies, further discussion on the equivalent resistance and capacitance of the device must be discussed.

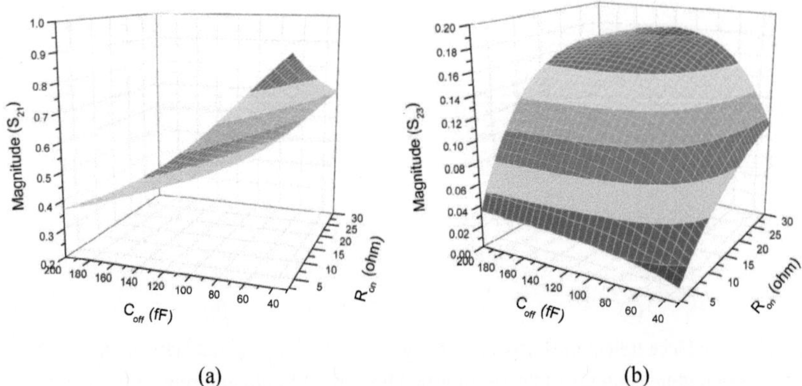

(a)   (b)

Figure 2.18: Simulated magnitude of (a) S_{21} and (b) S_{23} in terms of R_{on} and C_{off}.

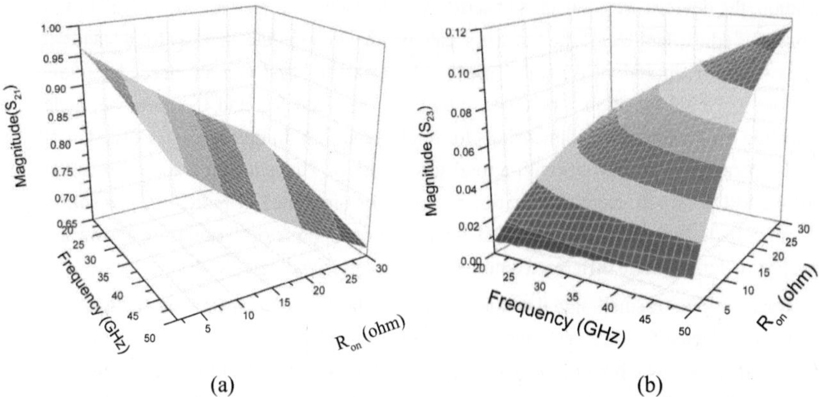

(a)   (b)

Figure 2.19: Simulated magnitude of (a) S_{21} and (b) S_{23} in terms of operating frequency and R_{on} with a fixed C_{off} of 30 fF.

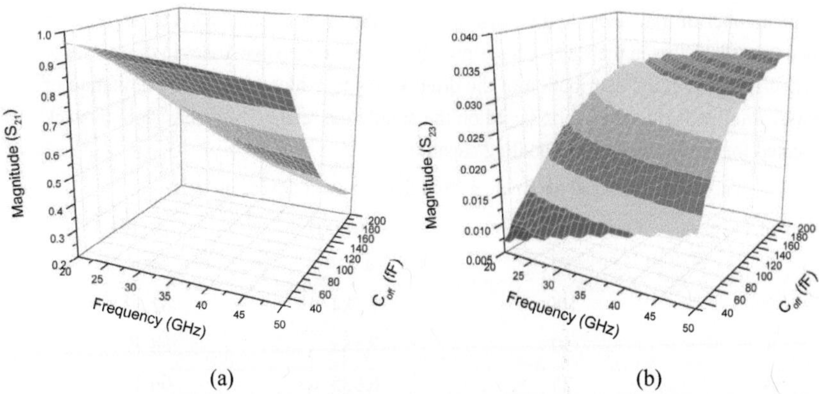

(a)   (b)

Figure 2.20: Simulated magnitude of (a) S_{21} and (b) S_{23} in terms of operating frequency and C_{off} with a fixed R_{on} of 2 ohm.

In order to improve the power handling capability of RF switches, stacked-FET configuration have been utilized in various of millimeter-wave switch designs [44], [51] and [56]. The stacked-FET configuration has been commonly used as the power stage in power amplifiers [57]-[59]. Technically, such a topology has the advantage of high output impedance since the devices are connected in series both DC- and RF-wise. For a SPDT RF switch using series-shunt configuration, the drain-to-gate breakdown of the series devices is one of the critical reasons limiting the overall power handling capability. Referring to equation (2.1)-(2.3), the stacked-FET configuration can effectively improve the power handling capability of RF switches as a higher voltage swing is allowed at high level input power. However, based on the evaluation from Fig. 2.18 to Fig. 2.20, a low equivalent R_{on} and low equivalent C_{off} are necessary for achieving simultaneous low insertion loss and high isolation. For stacked-FET configuration, the devices are connected in series, which leads to an increase of equivalent R_{on} and decrease of equivalent C_{off} if the transistor gate widths are kept the same as for a standard transistor switch configuration. As stated previously, small R_{on} and C_{off} are needed for low insertion loss and high isolation. Therefore, a trade-off between R_{on} and C_{off} must be made while selecting the proper device dimensions for implementation. In this case, an extraction of R_{on} and C_{off} parameters of devices with a total gate width of 200 μm was made prior to the design of the SPDT switch. Note that the device data were simulated from the 0.15-μm GaAs pHEMT process-design-kit (PDK) provided by WIN Semiconductors. Table 2.1 shows the extracted parameters of three different combinations of fingers and gated width.

With the consideration of small-signal characteristics and large-signal performance, the 4×50 μm device was selected for further designs of the SPDT switch. The schematic simulating for the series-shunt SPDT switch with stacked-FET configuration is shown in Fig. 2.21. Referring to [57], the largest improvement in maximum output power was reported to be two transistors in stack comparing with the number of 3 to 8. Based on the statement, a simulation of the input P_{1dB} performance with different number of devices in stack is plotted in Fig. 2.22. It is obvious that, as the number of transistors in the stack increases, the improvement starts to degrade for more than two transistors in the stack. This is mainly due to the resistive losses in the transistor and phase difference in the distribution of the drain voltages, which leads to the limitation of ideal power improvement. Therefore, based on the simulation results, two transistors in stack were selected for improving power handling capability.

Table 2.1 Extracted parameters from the provided device models

Number of Fingers	Gate Width (μm)	R_{on} (ohm)	C_{off} (fF)
2	50	14.829	34.79
2	100	7.584	62.42
4	50	7.183	68.39
8	25	6.935	66.25

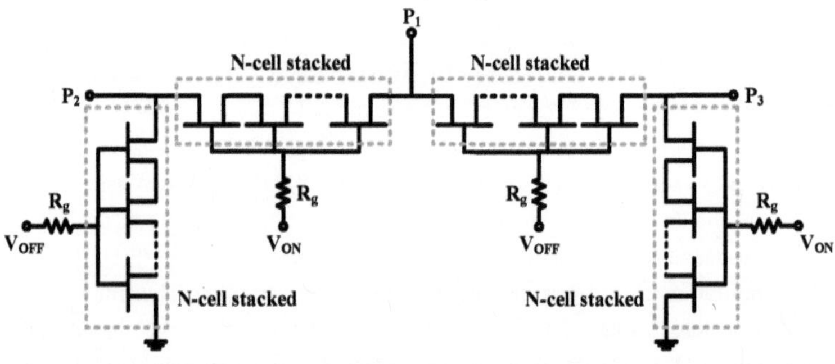

Figure 2.21: Schematic of series-shunt SPDT switch with N devices in stacked.

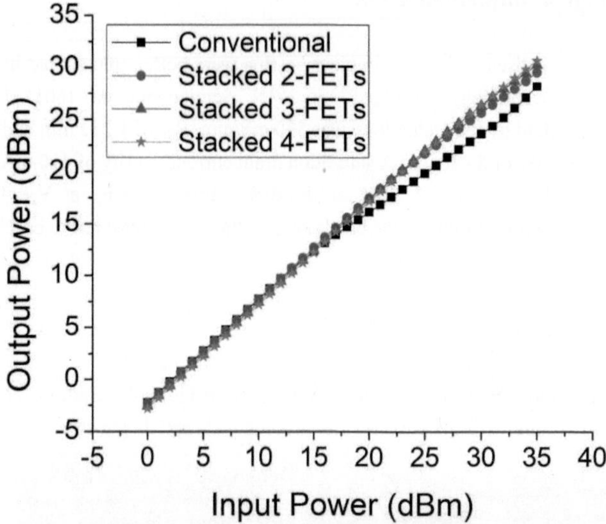

Figure 2.22: Simulated input P_{1dB} with different number of devices in stack.

Simulations of the small-signal characteristics with different number of shunt arms are performed in Fig. 2.23.

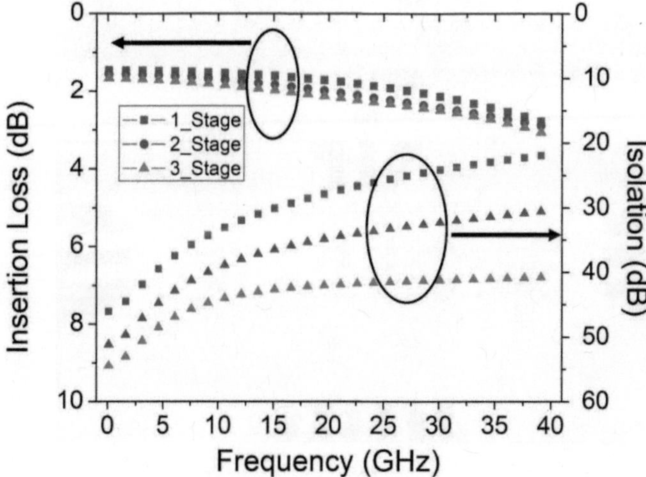

Figure 2.23: Simulated small-signal characteristics with different number of shunt arms.

As observed from Fig. 2.23, the number of shunt arms were determined to be two for achieving simultaneous reasonable insertion loss and isolation.

2.2.4. Technological implementation

The two SPDT switches discussed in the previous section were both implemented in the 0.15-μm GaAs pHEMT technology provided by WIN Semiconductors [60]. The peak transconductance specified by the technology was 570 mS/mm, R_{on} of 1.2 Ω·mm and a f_t of 70 GHz using a transistor size of 2×50 μm. A saturation drain current density of 410 mA/mm was measured at V_{gs} = 0.5V with V_{ds} = 1.5V, as the drain current density at V_{gs} = 0V was characterized as 80 mA/mm. Moreover, the breakdown voltage was measured to be 11.5V with a defined gate leakage current of 1 mA/mm.

2.2.5. Measurement Results

The photo of the two fabricated SPDT switches are shown in Fig. 2.25. Both of them have a dimension of 2 mm × 1 mm with all the on-wafer probing pads and dicing streets included.

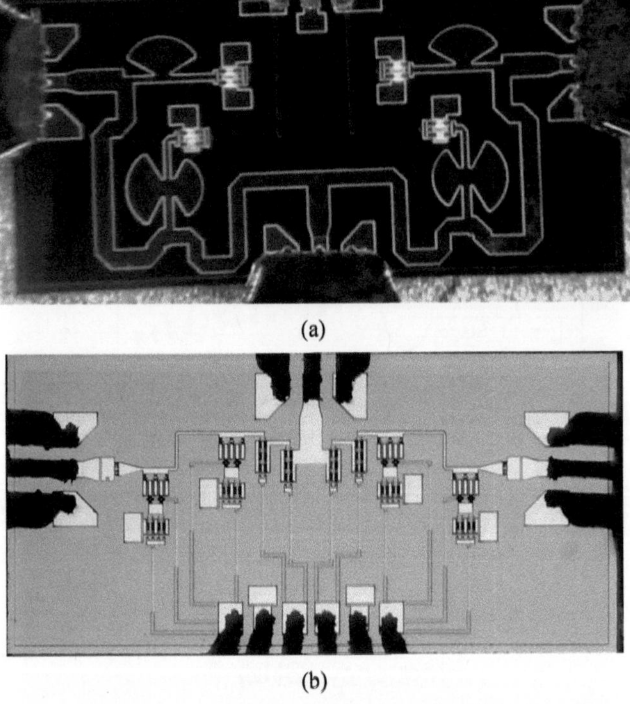

(a)

(b)

Figure 2.25: Chip photograph of (a) shunt type SPDT switch utilizing impedance transform method with radial stub and (b) series-shunt SPDT switch utilizing stacked-FET configuration.

The characterization of the two SPDT switches started with the small-signal performance. Fig. 2.6 shows the measurement setup for small-signal characterization using the 4-port Keysight N5227B PNA network analyzer. The measured small-signal characteristics are plotted in Fig. 2.27, where simulation results are included for comparison.

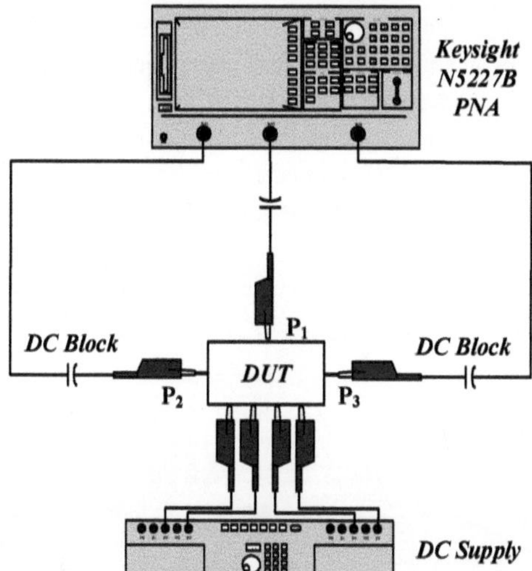

Figure 2.26: Measurement setup for small-signal characterization.

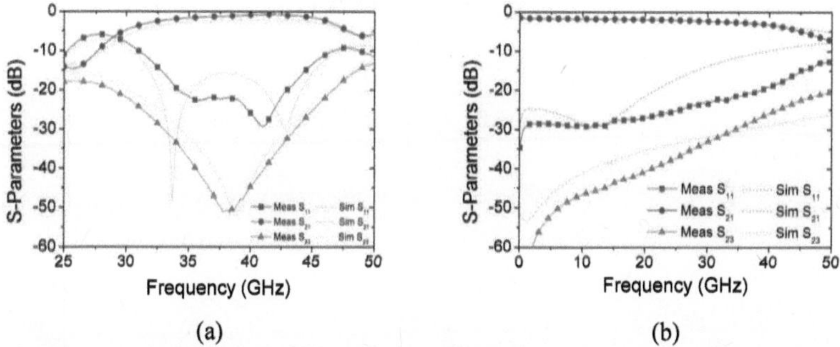

Figure 2.27: Small-signal performance of (a) SPDT switch utilizing impedance transform method with radial stub and (b) series-shunt SPDT switch utilizing stacked-FET configuration.

As shown in Fig. 2.27(a), the shunt-type SPDT switch utilizing the impedance transform method with radial stub exhibits an input return loss greater than 16 dB, insertion loss less than 2.5 dB and an isolation better than 30 dB across 33-44 GHz. For the series-shunt SPDT switch utilizing stacked-FET configuration, the input return loss was measured to be greater than 20 dB, insertion loss less than 3.2 dB with an isolation better than 27 dB from 8-40 GHz. With the shunt-type topology, the operating frequency is normally limited by the quarter-wavelength transmission line for separating the two branches. As for the series-shunt type switch, as the operating bandwidth becomes wider comparing with the shunt-type topology in general. This is mainly attributed to the replacement of quarter-wavelength transmission line with the series transistor.

The measurement setup for large-signal characterization is plotted in Fig. 2.28. The power was injected to P_2 of both switches and received from P_1 (Common input). Such arrangement was because that for the most cases, high power will be coming from a PA connected to one of the switch outputs in a real system. The measured output power for the two switches as a function of input power are plotted in Fig. 2.29. The two figures are plotted in different scale due to the different power handling capability. To avoid over-stress of the RF switch, both of the input power level was only measured up to the gate current level of 1 mA. The measured P_{1dB} at 38 GHz for SPDT switch with radial stub was 25 dBm and 34.9 dBm for the series-shunt SPDT with stacked-FET configuration.

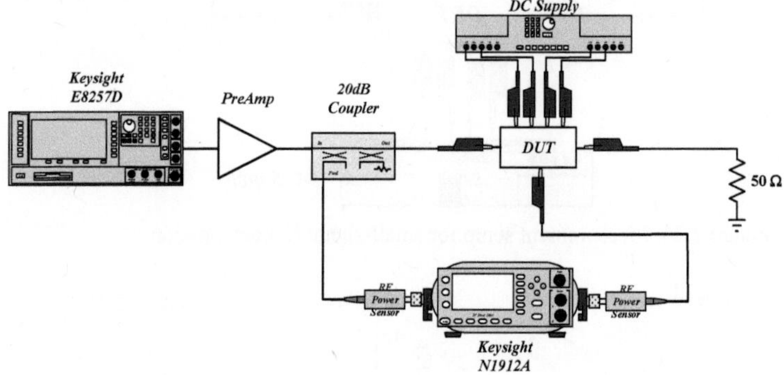

Figure 2.28: Measurement setup for large-signal characterization.

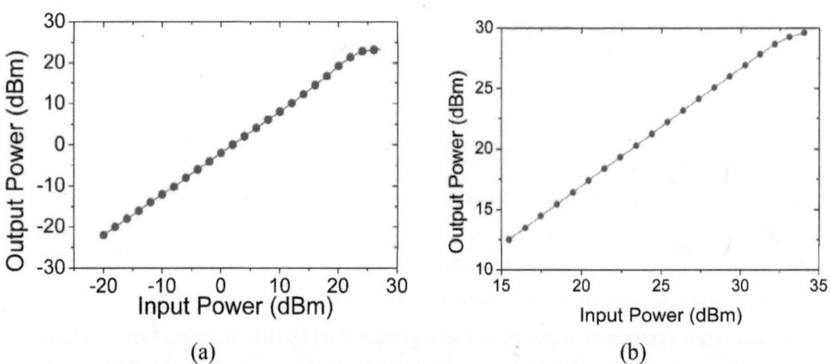

Figure 2.29: Large-signal performance of (a) SPDT switch utilizing impedance transform method with radial stub and (b) series-shunt SPDT switch utilizing stacked-FET configuration.

Fig. 2.30 plots the measured input P_{1dB} across the frequency of interest for both SPDT switches, where the X-axis was plotted as a function of the operating frequencies. Thus, Fig. 2.30 (a) is plotted from 33-41 GHz and Fig. 2.30 (b) is plotted from 26-40 GHz. The SPDT switch using impedance transform concept with implementation of radial stub was measured to have greater than 24 dBm input P_{1dB} from 34 to 40 GHz; as for the series-shunt SPDT switch with stacked-FET configuration, the measured input P_{1dB} was larger than 33.8 dBm from 26 to 40 GHz.

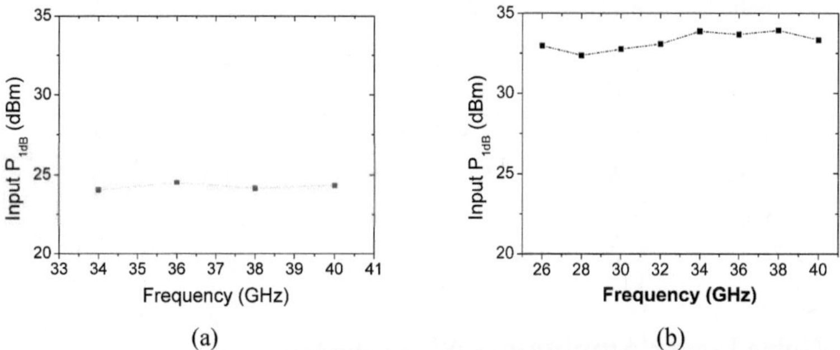

(a)   (b)

Figure 2.30: Measured input P_{1dB} of (a) SPDT switch utilizing impedance transform method with radial stub and (b) series-shunt SPDT switch utilizing stacked-FET configuration across frequency of interest.

In Table 2.2, the experimental results of the two SPDT switches are compared with previously published works. Note that RL represents return loss, IL stands for insertion loss and Iso for isolation. From the table of comparison, it is observed that the impedance transform SPDT switch with radial stub can effectively expand the operating bandwidth. It also maintains a reasonable power handling capability. For this specific case, the results are compared with [55] as a similar configuration was adopted for the SPDT switch design. The proposed topology using radial-stub as the matching component can truly improve the operating bandwidth with the experimental results. As for the series-shunt SPDT switch utilizing the stacked-FET configuration, the input P_{1dB} outperforms most of the reported works with a comparable operating bandwidth, reasonable insertion loss and isolation. To date, the reported series-shunt SPDT switch shows the greatest input P_{1dB} of 35 dBm at Ka-band frequencies. For the comparison between the shunt and series-shunt type SPDT switch discussed in this section, the first difference between them will be the operating frequency. The the utilization of the series-shunt topology, the SPDT switch exhibits a larger operating bandwidth comparing with the shunt type switch. The major difference between them will be the power handling capability. As the series-shunt type switch is characterized by the stacked-FET configuration to improve tolerance of the output voltage swing at the drain node of the transistors, an obvious difference of 10 dB was evaluated from the measurement results. The experimental results also revealed that with proper derivation of the intrinsic parameters of the applied transistors, a comparable insertion loss and isolation could be extracted by the theoretical analysis.

Table 2.2 Table of comparison between proposed SPDTs with published works

Reference	Frequency (GHz)	Input RL (dB)	IL. (dB)	Iso. (dB)	Input P_{1dB} (dBm)	Area (mm^2)
Radial-stub	33-44	> 16	< 2.5	> 30	> 24	2
Series-shunt stacked	8-40	> 20	< 3.2	> 27	> 33.8	2
[44]	35-70	> 15	< 3	> 40	20.2	0.96
[51]	22-26	> 10	< 2.5	> 44	36	3.12
[53]	25-30	> 14.2	< 1.1	> 27	31.8	0.043
[55]	38-43	> 15	< 2	> 30	N/A	2
[56]	15-35	> 10	< 3	> 25	26	2

2.3. Low-Noise Power Amplifier (LNPA) Design

As stated at the beginning of this chapter, the main idea combining low-noise amplifier and power amplifier into an LNPA was to reduce the dimension and complexity of the T/R module. In this section, the concept and design of the LNPA will be discussed.

2.3.1. Concept of LNPA

The latest implementation of an LNPA was reported in [61] at 2014. The single stage LNPA was proposed using a stacked-FET configuration to achieve simultaneous high gain and low noise characteristics across 4-15 GHz. Also, the output power density of the MMIC was improved with the utilization of a stacked-FET configuration. For multi-stage LNPA designs, the first stage must maintain low-noise performance, whereas the power stages need to to generate an output power suitable for driving the T/R module.

The design of low-noise amplifiers at millimeter-wave frequencies was widely reported in recent years. [62] proposed a low-noise amplifier in InP HEMT technology using high-pass reactive matching circuits with resistive-feedback network to simultaneously achieve low-noise and broadband characteristics. The block diagram of the five-stage low-noise amplifier was given as Fig. 2.30. It could be divided into the reactive matching part (for the first two transistor amplifiers) and resistive feedback part (for the last three transistor amplifiers). Matching loss induced by the matching circuits is considered as a challenge for millimeter-wave amplifiers. To reduce the matching loss, a thin-film microstrip line was applied to the reactive matching network for low matching loss while yielding low noise figure. Source degeneration was adopted for noise matching purpose to bring the optimum noise impedance (Γ_{opt}) and conjugate of input impedance (S_{11}^{*}) together for simultaneous matching of minimum noise figure and maximum transmission of the signal resulting in a high power gain. As the reactive matching circuit limited the maximum available gain of the devices, the resistive matching circuit helped to compensate the gain profile to achieve a flat gain over the frequency of interest. The bandwidth for high gain amplifier design was often considered challenging as the MAG and MSG were limited by the device characteristics. Therefore, the LNA presented in [62] uses a resistive feedback network for broadband matching purpose throughout the last three stages of the amplifiers. The gain profiles of the first two stages with reactive matching and the three-stage gain amplifier with resistive feedback topology were shown in Fig. 2.30 to form the flat gain profile across the frequency of interest. As shown in the figure, the two-stage LNA could focus on the low-noise matching to achieve minimum noise figure, whereas the third-stage gain amplifier focusses on the gain performance at higher frequencies with a limited contribution on total noise figure.

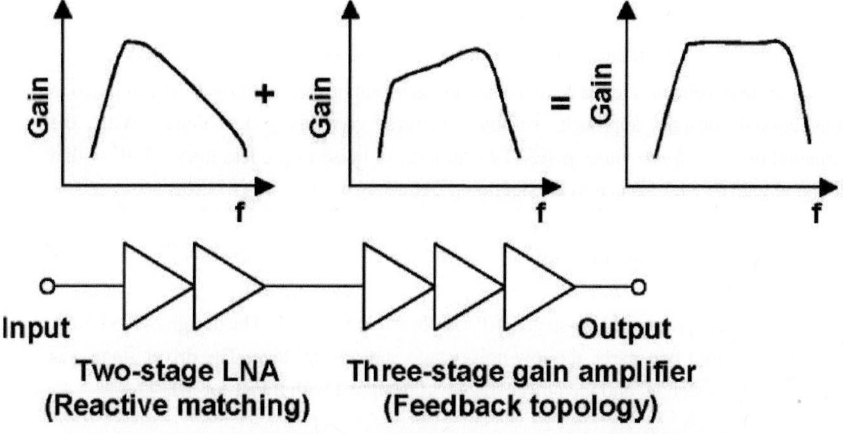

Figure 2.30: Block diagram of the five-stage low-noise amplifier [62].

The InP-based low-noise amplifier was reported with a high gain of 36 dB and a noise figure of 1.9 dB with the implementation of thin-film microstrip line for matching loss reduction comparing with conventional transmission line matching approach, where a matching loss reduction of 0.4 dB was reported.

A forward combining technique for amplifier gain boosting at millimeter-wave frequencies was proposed in [63]. The respective amplifier circuit using forward combining technique is plotted in Fig. 2.31. As shown in the figure, the design splits off and amplifies the input signals by realizing an anti-phase situation at the drain and source node of the device. After passing through a +90° and a -90° phase shifter between the drain and source nodes respectively, the signal is in phase again and is combined at the output:

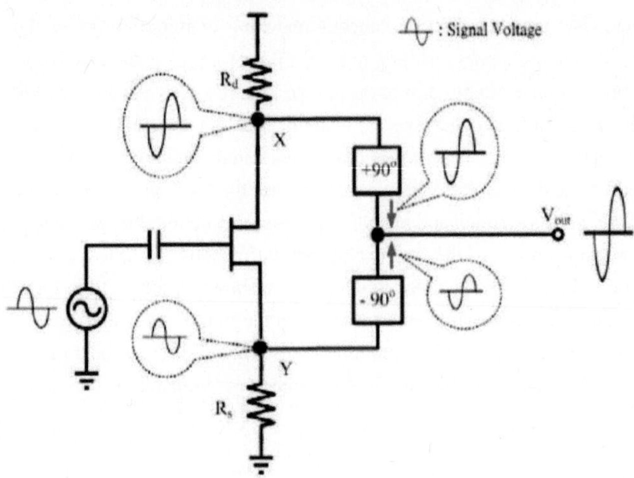

Figure 2.31: Schematic of amplifier using forward combining technique [63].

With such an arrangement, the maximum available gain can be increased by 3-dB compared to a conventional design approach without forward combining technique. With the implementation in 0.15-µm GaAs pHEMT technology, a noise figure less than 3.3 dB with a small-signal gain of 14.2 dB across the operating frequency of 29 to 43 GHz was achieved.

2.3.2. Analysis and Simulations

The schematic of the proposed two-stage LNPA is shown in Fig. 2.32. The design of the LNPA can be separated into two parts, the low-noise stage and power stage. The driver stage was designed with transistor M_1 and the power stage was designed with transistor M_2.

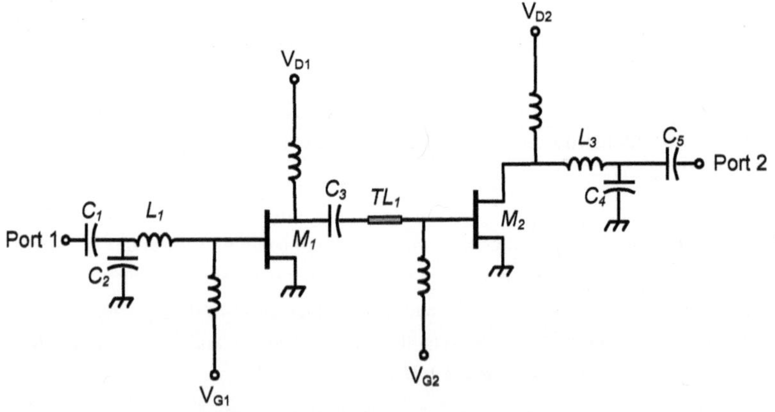

Figure 2.32: Schematic of proposed LNPA.

The design starts with the design of the low-noise stage. For a multi-stage amplifier, the overall noise figure can be expressed as equation (2.6).

$$F = F_1 + \frac{F_2 - 1}{G_1} + \frac{F_3 - 3}{G_1 G_2} + \cdots + \frac{F_N - 1}{G_1 G_2 \cdots G_{N-1}}$$

(2.6)

Based on the equation shown above, noise figure of the overall amplifier is mainly attributed to the first stage.

Prior to the design of the low-noise driver stage, an analysis of the Γ_{opt} and S_{11}^* is done using the small-signal equivalent circuit model as shown in Fig. 2.33. The input impedance and Γ_{opt} of the GaN HEMT can be derived according to equation (2.7)-(2.8). Note that C_{ds} and R_{ds} are included in Z_L for derivation.

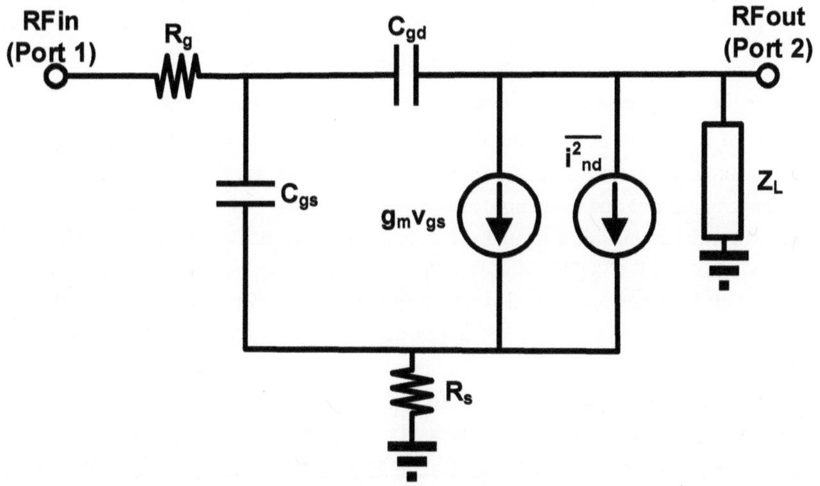

Figure 2.33: Small-signal equivalent model of the GaN HEMT device.

$$Z_{in} = \left(j\omega C_{gs} + j\omega C_{gd} \frac{1 + g_m Z_L}{1 + j\omega C_{gd} Z_L} \right)^{-1}$$

(2.7)

$$\Gamma_{opt} = \frac{1}{\omega \left(C_{gs} + C_{gd} \right)} \sqrt{\frac{2}{3}} R_g + \frac{1}{j\omega \left(C_{gs} + C_{gd} \right)}$$

(2.8)

The applied device for the first stage of the LNPA has a dimension of 2×50 μm. A noise measurement was made to characterize the minimum noise figure and Γ_{opt} of the device. The measured minimum noise figure (NF_{min}) is plotted in Fig. 2.34. To achieve NF_{min} of 1.9 dB at 38 GHz, the source impedance for the 1st transistor must be matched to the Γ_{opt} of $(9.5 + j\,13)$ Ω. From simulation, we can extract the S_{11}^* of the device as $(14.6 - j\,18.9)$ Ω, which allows a maximum power gain of 8.4 dB at 38 GHz.

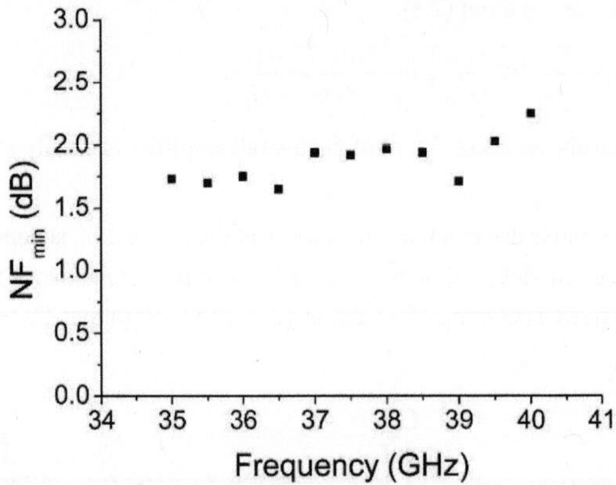

Figure 2.34: Measured minimum noise figure of the 2×50 µm device.

The schematic of driver stage design is shown in Fig. 2.35. A load-pull simulation targeting for the maximum output power was made using Keysight Advanced Design System (ADS), as the optimum source impedance and load impedance were extracted as $(14.1 - j\ 28)\ \Omega$ and $(25 + j\ 31.6)\ \Omega$. Within such termination, a simulated gain of 7.6 dB and saturated output power (P_{sat}) of 23.8 dBm was achieved at 38 GHz. However, the source impedance may not be an optimum termination for the LNA part and may therefore degrade the noise performance of the LNPA. Therefore, the final selection of the source impedance was set to $(8.6 + j\ 5)\ \Omega$ and a load impedance of $(25 + j\ 31.6)\ \Omega$ after trading-off for the gain and output power performance. These termination lead to a small-signal gain of 7.2 dB and P_{sat} of 22.5 dBm. The input matching network was then designed after the determination of targeted source impedance. A 1-step LC matching network was then designed to transform the selected impedance to the 50-ohm load termination impedance for further measurement and integration purpose.

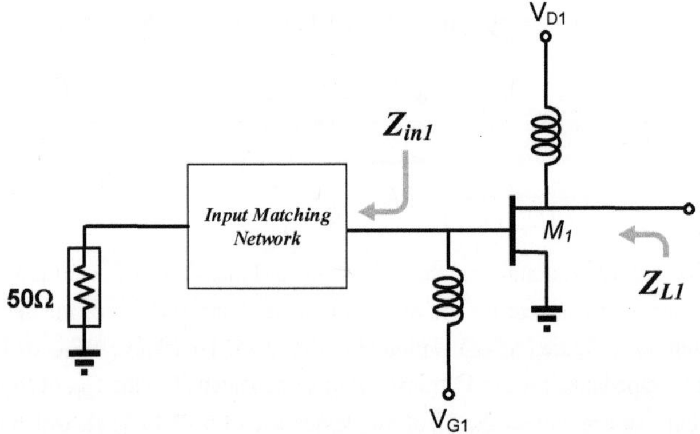

Figure 2.35: Schematic of driver stage amplifier with input matching network.

The design procedure then considers the power stage amplifier, the schematic is shown in Fig. 2.36. To maintain the drivability of the LNPA, a 2-to-1 ratio was determined for the devices, as a 4 × 50 μm device was selected for the power stage. Such an arrangement takes into account the drivability of the driver stage to achieve a reasonable efficiency when both stages were operating at the peak region. A load-pull simulation using Keysight ADS was also performed for the power stage design. An optimum source impedance of $(8.7 - j\,6.2)\ \Omega$ and an optimum load impedance of $(16 + j\,35.2)\ \Omega$ was extracted for power matching conditions. Without the consideration of matching loss, the terminated impedances would lead to a P_{sat} of 27.5 dBm and small-signal gain of 6.2 dB at 38 GHz. In order to improve the gain of the LNPA, the source impedance of the power stage is desired to be matched to $(9.7 - j\,12.7)\ \Omega$ based on the simulation results. The terminated impedance has lead to a small-signal gain of 6.8 dB and a P_{sat} of 27 dBm.

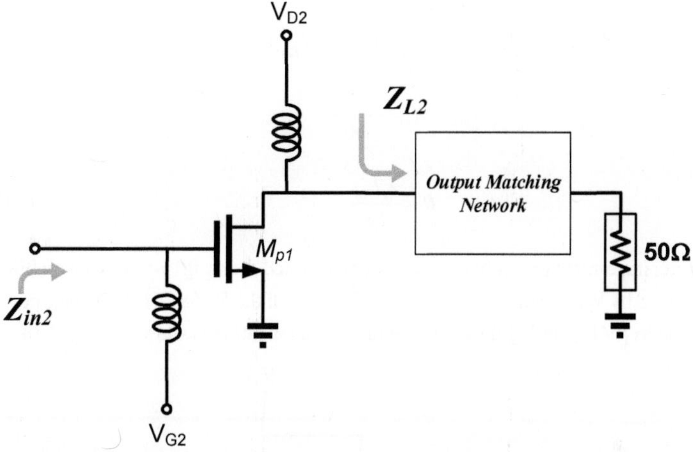

Figure 2.36: Schematic of power stage amplifier with output matching network.

Interstage matching usually decreases MAG of the complete amplifier. In order to keep this effect as small as possible, a matching topology with low loss must be applied. In [64], a low-Q interstage matching topology was reported with a 1.5-dB reduction in matching loss and an overall size reduction of 67% at 54 GHz. Instead of the quality factor defined for capacitors or inductances, the Q-factor mentioned in this reference paper was about the real and imaginary impedance ratio located on the Smith Chart. Take a simple matching topology as an example, as the route of matching passes through a high Q-factor region, the operating bandwidth tends to narrow; as for matching network passing through only low Q-factor region, the operating bandwidth will be larger. Moreover, this can also be referred to the matching loss induced by the matching network, where a lower matching loss was normally induced by the one passing through low Q-factor region. The design equations for the interstage matching was given as (2.9)-(2.12), the schematic is shown in Fig. 2.37.

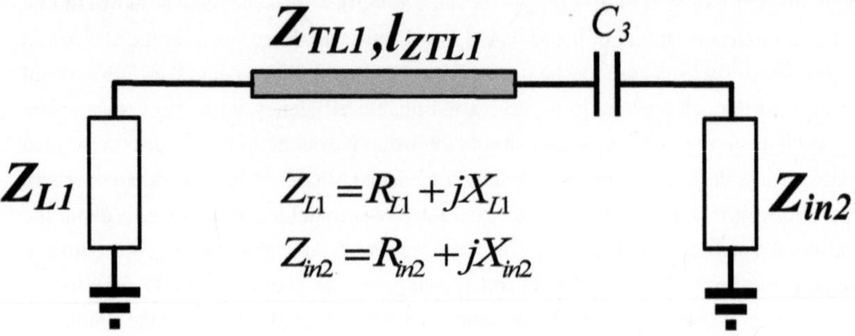

Figure 2.37: Schematic of low-Q interstage matching topology for loss reduction.

$$R_{L1}\left(Z_{TL1}+R_{in2}\tan\theta\right)+X_{L1}X'_{in2}\tan\theta = Z_{TL1}\left(Z_{TL1}\tan\theta+R_{in2}\right) \tag{2.9}$$

$$R_{L1}X'_{in2}\tan\theta - R_{in2}X_{L1}\tan\theta = Z_{TL1}X'_{in2} \tag{2.10}$$

$$X'_{in2} = X_{in2} - \frac{1}{\omega C_3} \tag{2.11}$$

$$\theta = \frac{\lambda_g l_{TL1}}{2\pi} \tag{2.12}$$

The low-loss interstage matching topology was demonstrated using 0.15-µm GaAs pHEMT technology provided by WIN Semiconductor. The schematic of the 52-58 GHz two-stage power amplifier reported in [64] targeting for inter-satellite link application is shown as Fig. 2.38.

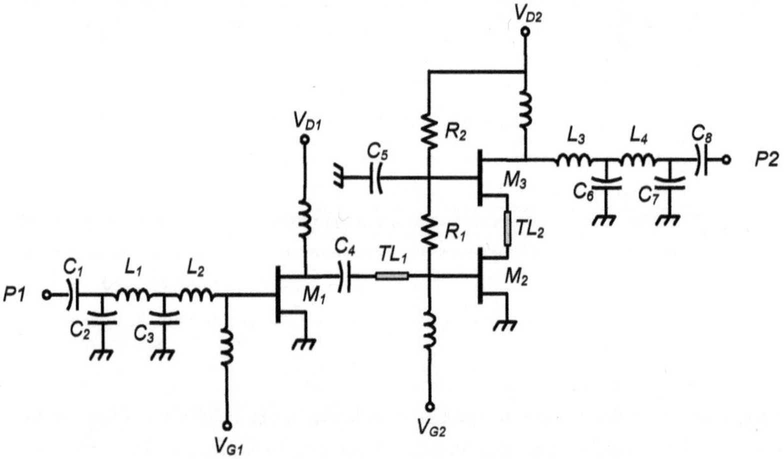

Figure 2.38: Schematic of 52-58 GHz power amplifier for space application.

In order to improve the output performance, a stacked-FET configuration was utilized for the power stage design. The resistive voltage divide was designed to properly distribute the supplied voltage to the two transistors M_2 and M_3. Table 2.3 summarizes the comparison of matching loss induced by different matching topologies in the paper for interstage matching. Note that all the matching networks were designed to match the operating bandwidth from 52-58 GHz, with an input/output return loss greater than 10 dB for fairness of comparison. Obviously, a maximum reduction of 1.5 dB in the matching loss was achieved by the low-loss interstage matching. Furthermore, such approach also reduced the dimension of the interstage matching network by 67% comparing with the 2-section distributed LC network.

Table 2.3. Matching loss induced by different matching topologies

Frequency (GHz)	Proposed Topology (dB)	1-section Lumped LC (dB)	1-section Distributed LC (dB)	2-section Lumped LC (dB)	2-section Distributed LC (dB)
52	-1.61	-3.80	-3.16	-3.58	-2.80
53	-1.61	-3.79	-3.20	-3.57	-2.95
54	-1.62	-3.78	-3.29	-3.56	-3.10
55	-1.63	-3.76	-3.40	-3.55	-3.27
56	-1.67	-3.75	-3.51	-3.54	-3.46
57	-1.71	-3.74	-3.64	-3.54	-3.60
58	-1.76	-3.72	-3.79	-3.53	-3.71

With the implementation of the low-loss interstage matching network, a physical length of 310 µm with a characteristic impedance of 45 ohm was derived for the series transmission line, as the DC blocking capacitance was set to be 750 fF. With such combination, a low interstage matching loss of 0.7 dB was achieved to maintain the gain of the LNPA. A simulated small-signal gain of 13 dB, P_{sat} of 27 dBm and a peak power-added-efficiency (PAE) of 15% was achieved at 38 GHz.

2.3.3. Technological implementation

The two-stage LNPA was implemented in the 0.15-µm GaN-on-SiC technology provided by FBH, where the technological device structure is shown in Fig. 2.39. The $Al_{0.28}Ga_{0.72}N$/GaN HEMT structures for Ka-band microwave applications were grown on 4 inch SiC wafers. Starting with the deposition of 50 nm AlN as a nucleation layer, a 1.9 µm uid GaN buffer followed by an Fe doped GaN buffer layer was grown. The GaN channel layer is designed such that the Fe-concentration in the channel layer decays below SIMS detection limit at the vertical position of the two-dimensional electron gas (2DEG). This avoids excessive buffer lagging of the devices during dynamic operation. Afterwards the growth of a less than 1 nm AlN interface layer followed by 10 nm $Al_{0.28}Ga_{0.72}N$ barrier layer follows. The fabrication of MMICs was accomplished using commercial NIKON i-line stepper lithography for defining all processing levels with the exception of those associated with the gate structures. The gates themselves rely on the so-called Ir sputter gate technology, a special technique developed at FBH [65] and were defined by electron beam lithography. The Ir sputtered gates are directly deposited in a slanted trench profile dry-etched into the first SiNx passivation layer. The sputtered gate structures have shown significantly reduced gate lagging and showed improved reliability data as compared to standard gate technologies. A very low ohmic contact resistance of 0.29 Ω·mm was achieved by using Ti/Al/Ni/Au/Pt metallization scheme annealed at 815°C.

The FBH 0.15-µm GaN-on-SiC technology showed further advantages in terms of high isolation by ion implantation and MIM capacitors with a high breakdown voltage by using SiN_x dielectric as passivation.

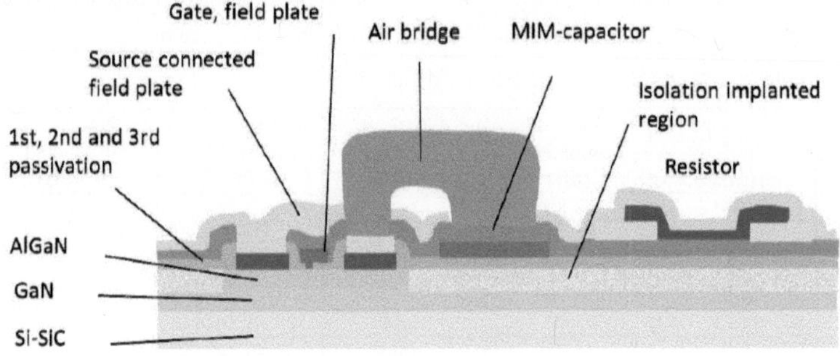

Figure 2.39: Device structure for FBH Ka-band MMIC process technology [65].

The 0.15-µm GaN-on-SiC coplanar MMIC process achieved a PAE of 50% at 10 GHz with a f_t of 60 GHz and f_{max} of 100 GHz extracted from the measurement results using 2 × 50 µm. The measured IV characteristics of a 2 × 50 µm device is plotted in Fig. 2.40, where a saturation drain current density of 750 mA/mm was measured at V_{gs} = 0V with V_{ds} = 10V. Moreover, with the defined gate leakage current of 1 mA/mm, the breakdown voltage specified by the technology was larger than 100V. The load-pull measurement results in Fig. 2.41 shows a maximum PAE of 27% with an output power density of 3.16 W/mm at 38 GHz.

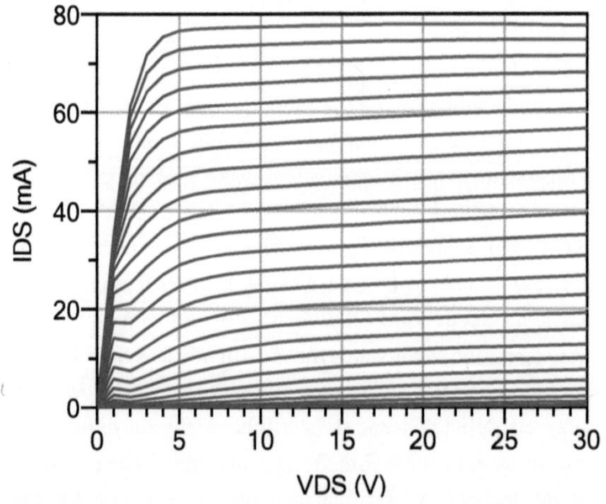

Figure 2.40: Measured ID-VD curve of a 2 × 50 µm device.

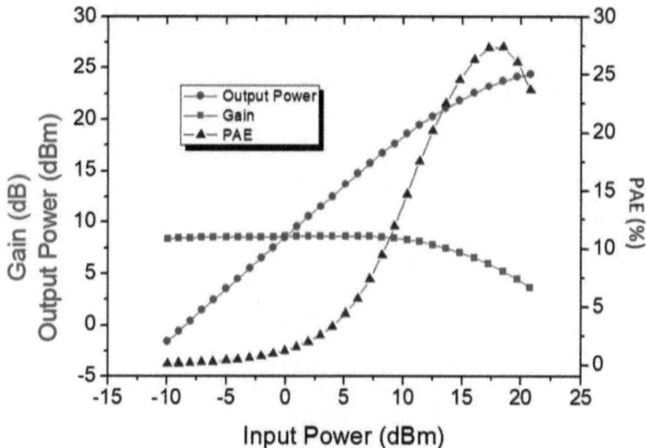

Figure 2.41: Load-pull measurement results of a 2 × 50 µm device biased at V_D = 18 V, V_G = -1.3 V.

2.3.4. Measurement Results

The experimental verification of the two-stage LNPA design will be presented in this section. Fig. 2.42 shows the photo of fabricated MMIC with a chip size of 2 mm × 2 mm including all the on-wafer testing pads and dicing street.

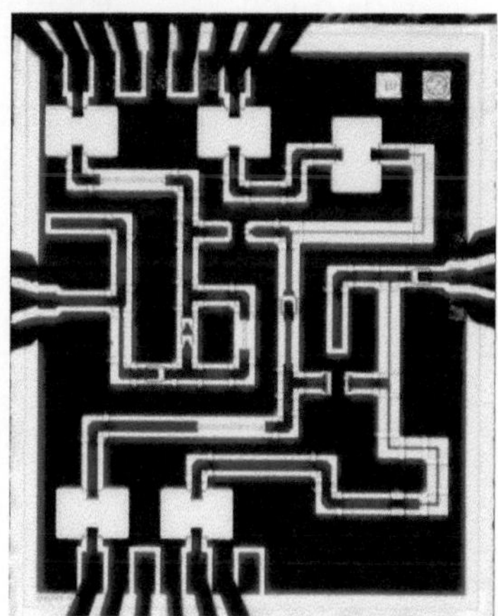

Figure 2.42: Fabricated chip photo of the two-stage LNPA according to Fig. 2.32.

The measurement setup for small-signal characterization is plotted in Fig. 2.43. Fig. 2.44 shows the measured results from 15-50 GHz, where the bias voltages were set as V_{D1} = 15 V, V_{G1} = -1.3 V, V_{D2} = 15 V and V_{G2} = -1.3 V. The corresponding currents of the driver and power stages were 12 mA and 21 mA, respectively. The simulated response of the two-stage LNPA is also included for comparison. The measured input return loss is greater than 5 dB from 36-40 GHz. This may be expected as the input of the transistor is noise-matched which usually is associated with a certain input reflection factor. Based on the measurement results, the degradation in the input return loss seemed not to affect the gain performance too much. The difference between simulation and measurement might be resulted in the design modeling issue and the setup of electromagnetic (EM) simulation for passive components. As for the output return loss exhibits a nice impedance match with better than 8 dB from 36-40 GHz. The LNPA delivered a small-signal gain of 14.3 dB, which was quite close to the maximum available gain, evidencing that the low-loss interstage matching topology has enhanced the overall gain.

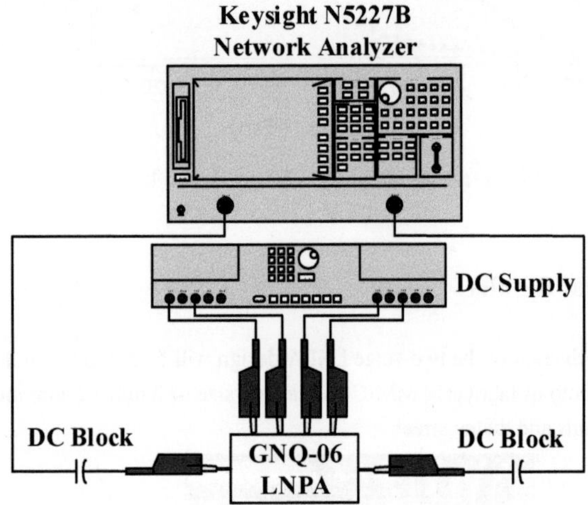

Figure 2.43: Measurement setup for the small-signal characterization.

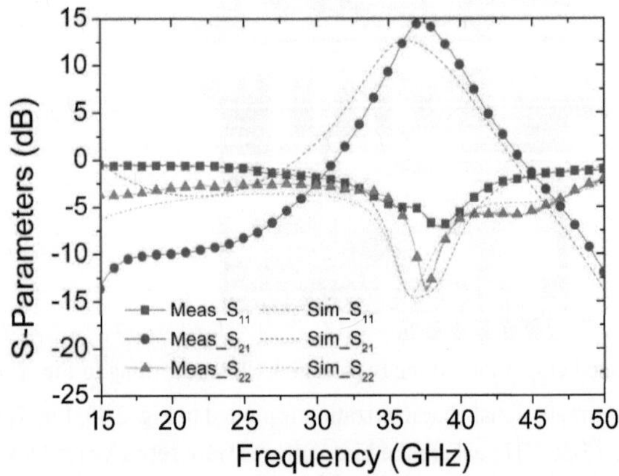

Figure 2.44: Small-signal performance of the two-stage LNPA.

Fig. 2.45 shows the large-signal measurement setup. Fig. 2.46 shows the measured gain and PAE versus output power at 38 GHz under CW mode excitation. As shown, the LNPA delivered a P_{1dB} of 21.66 dBm, P_{sat} of 25.52 dBm and a peak PAE of 16.27%. The measured P_{sat} and PAE as functions of frequencies are plotted in Fig. 2.47.

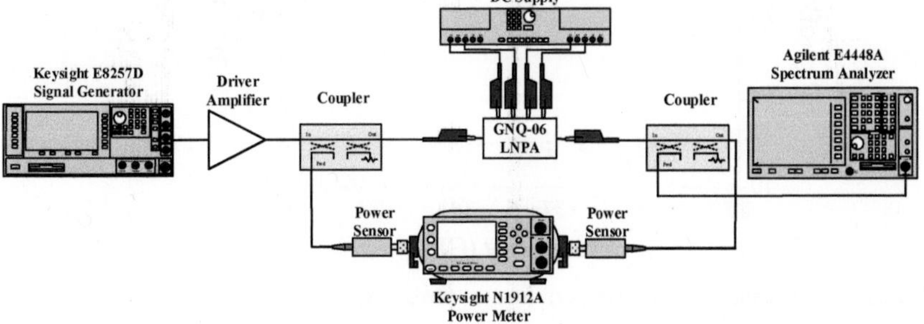

Figure 2.45: Measurement setup for the large-signal characterization.

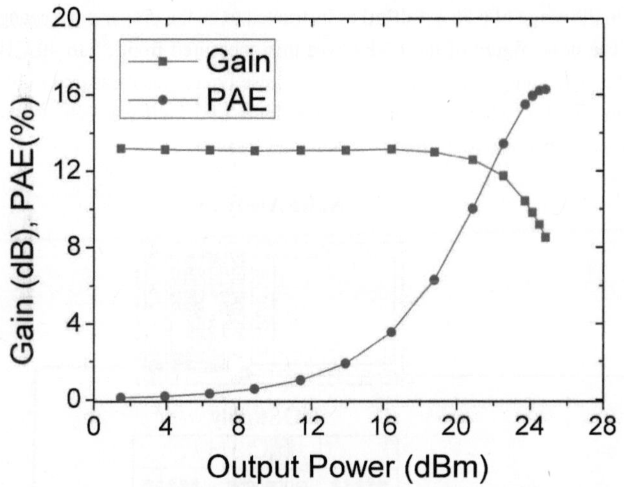

Figure 2.46: The measured gain and PAE versus output power at 38 GHz.

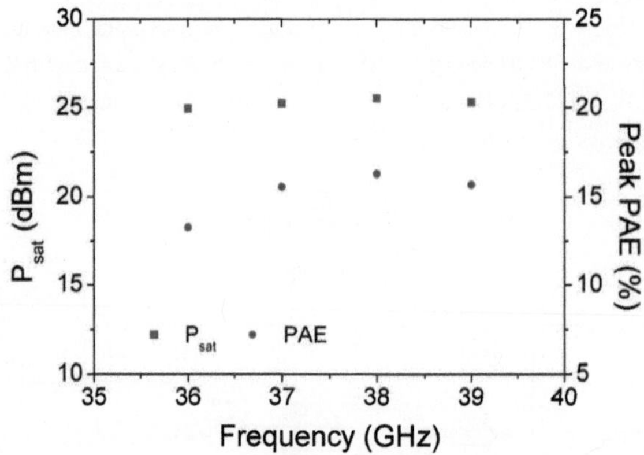

Figure 2.47: The measured power performance across frequency range of interest.

For an LNPA, the noise performance is important for Rx mode operation. Therefore, a noise measurement using the setup shown in Fig. 2.48 has been performed for characterization. The reference noise source was generated by a Keysight 346CK01 system. The output power of the noise source was tested under the suggested operating condition, where the power level was -53 dBm, which fell in the range of (-56 ± 4 dBm) as instructed to be the proper power range of measurement [66]. The noise figure of the LNPA was then measured from 35 to 40 GHz in frequency steps of 0.5 GHz and plotted in Fig. 2.47. The noise figure was measured to be less than 4 dB from 36 to 39 GHz, where a minimum noise figure of 3.2 dB was achieved at 36 GHz, exhibiting a reasonable noise performance at Ka-band frequencies.

Figure 2.47: Measurement setup of noise figure characterization.

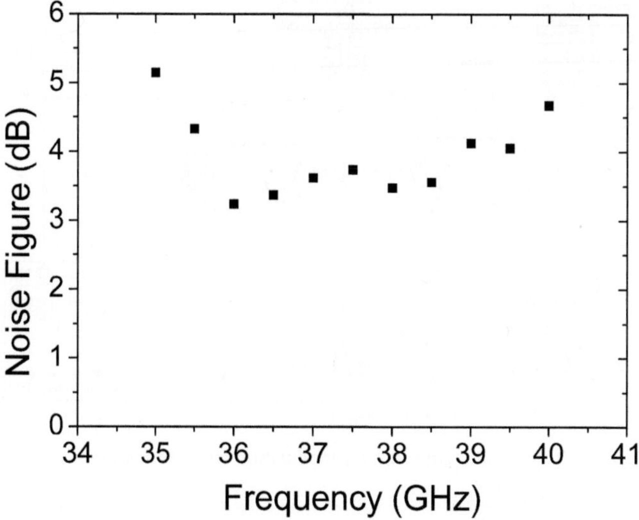

Figure 2.48: Measured noise figure across frequency of interest.

While targeting for further integration with a T/R module, characterization using modulated signals must be made to evaluate the linearity of the LNPA. The measurement setup for modulation characterization is shown in Fig. 2.49. The 64-QAM with PRBS11 signal and 0.35 roll-off factor (α) of the square-root-raised-cosine pulse-shaping filter (SRRC) were used to test the LNPA. Fig. 2.50 shows the measured error vector magnitude (EVM) of 64-QAM with 1.5 Gbps data rate versus different average output power at 38 GHz. The measured EVM was under -29.07 dB at the average output power level of 22.53 dBm at 38 GHz, exhibiting a relatively nice linearity with the high-speed modulation scheme. Based on the announced specification for 3GPP 5G NR FR2 carrier, the regulated EVM level for the output signal was set at -21.94 dB for 64-QAM modulation scheme. From Fig. 2.50, the EVM at linear region showed a low EVM level as the power gain of the LNPA did not degrade too much. According to the measurement results, the LNPA showed a nice EVM results for operation up to the output power level of 24.11 dBm. This meets the specification announced for 5G NR FR2 64-QAM applications, which is less than 8% EVM. Fig. 2.51 shows the measured constellation and the corresponding spectrum of 64-QAM with 1.5 Gbps data rate at the average output power of 22.53 dBm at 38 GHz. For the constellation, the signal distortion began with the amplitude distortion (AM-AM distortion). After passing through the medium power region, the phase difference of the signal started to increase, which led to the degradation in AM-PM performance. Therefore, the measurement results of the symbols in the constellation can then be converted to the EVM parameter.

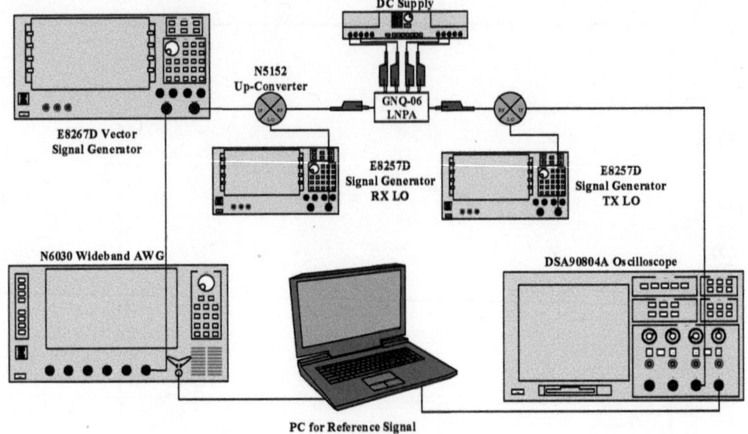

Figure 2.49: Measurement setup for modulation characterization.

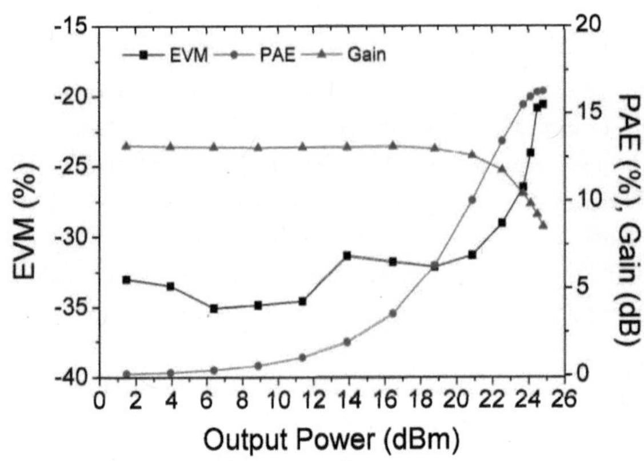

Figure 2.50: The measured EVM versus average output power at 38 GHz.

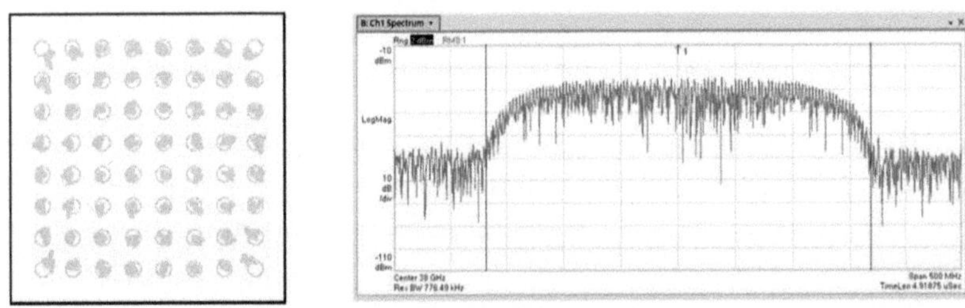

Figure 2.51: Measured constellation and spectrum of 64-QAM with 1.5 Gbps data rate at the average output power of 22.53 dBm at 38 GHz.

To compare the performance of the LNPA with previously published works, this may have to be separated into two parts, low-noise amplifiers and power amplifiers. Table 2.4 summarizes the performance of the LNPA and previously published LNA works. The LNAs reported in [67] and [70]-[71] concentrating on similar operating bandwidth compared to the proposed two-stage LNPA showed a worse noise performance. Moreover, the output P_{1dB} level of the designed two-stage LNPA also outperforms these referenced works. As for the LNAs proposed in [68] and [69], the measurement results exhibited an extraordinary low noise performance. However, no detail discussion on the design approach and measurement setup were reported in the paper. The power consumption was also comparable although biasing at a high drain voltage of 18V. The experimental results reveal that the LNPA delivers a reasonable performance compared to published LNAs. For large-signal consideration, the LNPA shows a relatively high power consumption. This is mainly attributed to the high-power operation at the power stage, which requires a higher supplied voltage.

Another comparison with the published linear power amplifiers reported with different device technologies are summarized in Table 2.5. From the modulation measurement results, the proposed LNPA using FBH 0.15-μm GaN/SiC HEMT technology achieved only 2.2-dB back-off from P_{sat} to achieve an EVM less than -26.94 dB, as the CMOS power amplifiers required over 9-dB back-off and GaAs technology with 4.9-dB back-off. The measurement results prove that the LNPA is well suited for the integration of T/R module while fulfilling the requirements regulated by 3GPP with an EVM level lower than 8% under 64-QAM modulation scheme [85]. The PAE performance at linear power region is also showing a competitive results comparing with the reported works in the comparison table. Device technologies including CMOS and GaAs were reported to also show nice PAE at Ka-band frequencies. However, the peak PAE performance was normally measured near to the saturated level, where the linearity of the PA has been degraded a lot already. In the measured results of the two-stage GaN LNPA, the peak PAE was observed close to the saturation region with only a few back-off from the measured P_{sat}. The nice linearity performance may be due to the low source resistance induced by the utilization of Ir-sputterd gate configuration. Moreover, potential implementation of highly linear and high output power amplifier using GaN technology are promising for the future communication systems.

Table 2.4 Comparison table with the proposed LNPA and previously published LNAs.

Ref.	Tech.	Freq. (GHz)	Gain (dB)	NF (dB)	P_{1dB} (dBm)	P_{dc} (mW)	Area (mm²)
This work	0.15-μm GaN/SiC	36-40	14.3	3.42	21.66	510	4
[67]	0.15-μm GaN/SiC	27-31	12	3.9	N/A	560	4.08
[68]	0.1-μm GaN/Si	22-30	20	1.1	20.8	210	2.21
[69]	0.1-μm GaN/Si	18-56	18	2.2	15	1400	4.8
[70]	0.15-μm GaN/SiC	42-47	19	3.7	N/A	400	3.3
[71]	0.12-μm GaN/SiC	33-41	13	4	13	280	0.7

Table 2.5 Comparison table with the proposed LNPA and previously published PAs.

Reference	This Work	[72]	[73]	[74]
Technology	0.15-μm GaN/SiC	90-nm CMOS	28-nm CMOS	0.15-μm GaAs
Frequency (GHz)	38	34	30	28
Gain (dB)	13.2	12.6	15.7	12.6
P_{1dB} (dBm)	21.66	18.5	13.2	24.55
P_{sat} (dBm)	25.52	20.7	14	26.5
Peak PAE (%)	16.56	32.7	35.5	31
Modulation	64-QAM	64-QAM	64-QAM	64-QAM
P_{Linear}	23.35[a]	11.3[a]	4.2[b]	21.6[b]
PAE(%) @P_{Linear}	15.66	5.1	9	13.5
Backoff from P_{sat} (dB)	2.2	9.4	9.8	4.9
V_D (V)	15	1.2	1	4

[a]: defined at the Error-Vector-Magnitude (EVM) level of -27 dB

[b]: defined at the Error-Vector-Magnitude (EVM) level of -25 dB

2.4. Ka-band Dual-band Power Amplifier Design

As 5G communication systems are targeting for a higher data rate, a wider transmission bandwidth is needed to achieve the goal. Therefore, carrier aggregation (CA) technique is implemented for increasing the transmission bandwidth [75]-[77]. A successful implementation using non-contiguous carrier aggregation for high-speed 5G downlink with a throughput up to 10 Gbps was released by Nokia and Optus in April, 2021 [78]. Through the adoption of inter-band CA, a total channel bandwidth of 800 MHz at millimeter-wave frequencies was obtained from the n257 and n260 bands. Obviously, dual-band power amplifiers (PA) at millimeter-wave frequencies are essential for next-generation 5G and beyond 5G systems [79]. Moreover, power amplifiers with identical performance at both bands are preferred to guarantee the overall system performance.

2.4.1. Concept of dual-band power amplifier

Dual-band design techniques have been widely reported PA designs through the past years. The design techniques mainly focus on the implementation of matching topologies for achieving dual-band characteristics, such as concurrent matching. The concurrent configuration requires the design of biasing network with the resonating frequencies at the desired operation point. The major advantages were reported to be the compact size and no less degradation in the efficiency. In [80], an energy-efficient Ka/Q dual-band two-stage PA was reported with using the commercial 0.1-μm GaAs pHEMT technology. The dual-band matching networks for output, input and interstage design have been synthesized using design equations for specific operating frequencies. Fig. 2.52 shows the reported dual-band matching networks. Fig. 2.52 (a) was designed for the output matching as the structure in Fig. 2.52 (b) was applied for both input and interstage matching. Thus, P_1 would either be input termination or the source node of the power stage amplifier; as for P_2, the port would be source node of the driver stage and the power stage amplifier.

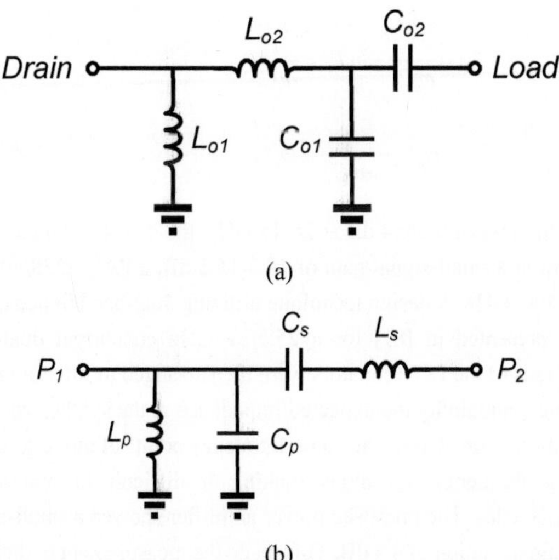

Figure 2.52: Dual-band matching network for (a) output, (b) input and interstage [80].

For the output network, a low insertion loss is required for achieving high efficiency. The shunt inductor L_{o1} was designed as the drain bias and the series capacitor C_{o2} for DC blocking purpose. As for L_{o2} and C_{o1}, these components were implemented in the matching network for impedance transformation at the targeted operating frequencies. As for the interstage and input matching network, it can be equivalent as Fig. 2.53 at ω_1 and ω_2.

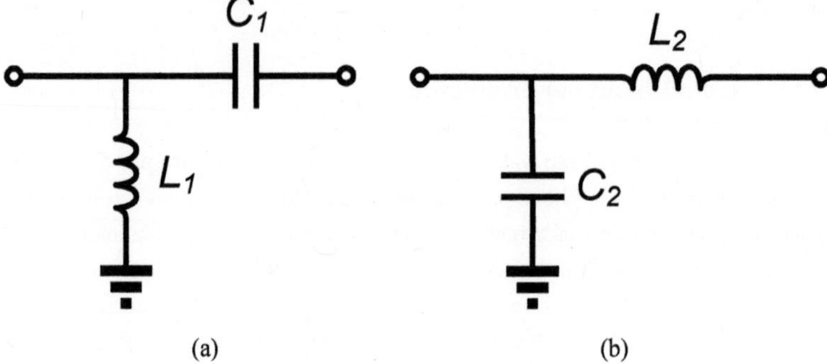

Figure 2.53: Equivalent circuit for interstage and input matching network at (a) ω_1 and (b) ω_2. The matching networks were synthesized using the given equations:

$$\frac{1}{\omega_1 L_p} - \omega_1 C_p = \frac{1}{\omega_1 L_1} \tag{2.13}$$

$$\omega_2 C_p - \frac{1}{\omega_2 L_p} = \omega_2 C_2 \tag{2.14}$$

$$\frac{1}{\omega_1 C_s} - \omega_1 L_s = \frac{1}{\omega_1 C_1} \tag{2.15}$$

$$\omega_2 L_s - \frac{1}{\omega_2 C_s} = \omega_2 L_2 \tag{2.16}$$

With a simulated matching loss of 0.6/0.4 dB at 28/45 GHz, the dual-band matching network, the power amplifier showed a small-signal gain of 19.3/15.5 dB, a PAE of 38/40 % and a P_{sat} of 22.5/22.7 dBm at 29.5/47 GHz. A design technique utilizing dual-band impedance matching filtering networks was presented in [81] for a 25.5/37 GHz concurrent dual-band power amplifier. The combinations of the LC resonators were then arranged to filter out the unwanted out-of-band signals while maintaining the expected impedance at the synthesized frequencies. Since the synthesis equations were derived at single frequency points of interest, the impedance generally shows strong frequency variations making it difficult to maintain identical performance at dual frequencies. The proposed power amplifier showed a small-signal gain of 21.4 and 17 dB with a gain deviation of 4.4 dB. This made the measurement in dual-band mode complicated as an external 4.4 dB attenuation was necessary to compensate the difference between the two frequencies. Fig. 2.54 shows the measured spectrum with dual-band mode operation at saturation, where the measurement setup is also included. For such arrangement, it might lead to the high complexity for further system level integration, as an external attenuator is required for proper carrier aggregation. Therefore, a dual-band PA with identical performance at the two operating frequency bands for carrier aggregation is highly desired for further system integration.

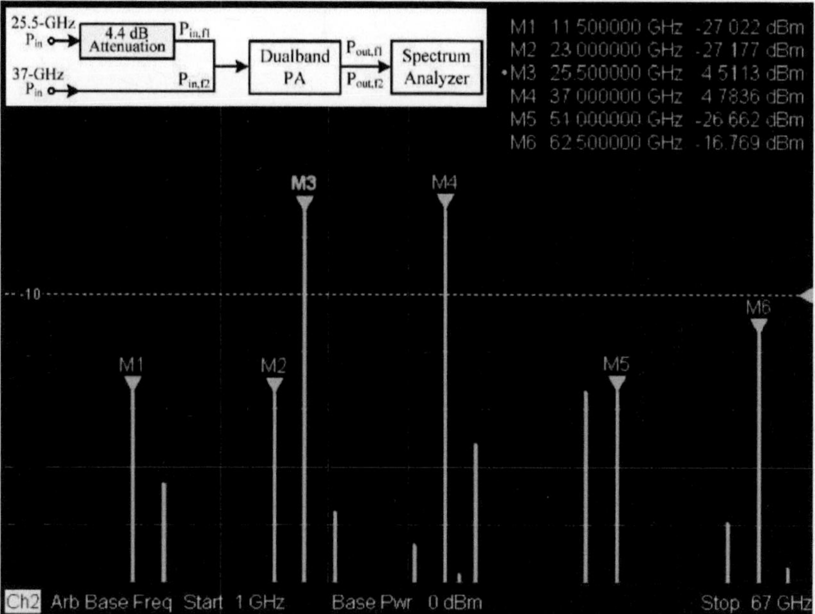

Figure 2.54. Measured output spectrum at saturation with 4.4 dB attenuation at 25.5 GHz [81].

2.4.2. Analysis and simulations

Fig. 2.55 shows the schematic of the dual-band power amplifier implemented with 0.15-μm GaAs pHEMT technology provided by WIN Semiconductor. The dual-band power amplifier utilized the stacked-FET configuration with adoption of resistive biasing network. The design parameters are given in Table 2.6. Note that the capacitances are in fF and inductances are in pH.

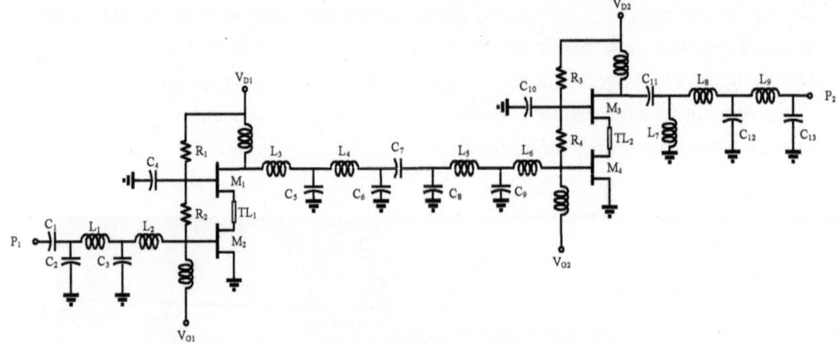

Figure 2.55. Schematic of the dual-band power amplifier.

Table 2.6 Design parameters of the dual-band power amplifier.

C_1	L_1	C_2	L_2	C_3	R_1	R_2	Z_{TL1}	l_{TL1}
214.05	194.61	75.05	73	303.3	600	500	77	150
M_1	M_2	L_3	C_4	C_5	L_4	C_6	L_5	C_7
4 x 75	4 x 75	360.68	150	123.4	289.14	40.24	164.97	144.29
L_6	C_8	R_3	R_4	Z_{TL2}	l_{TL2}	M_3	M_4	C_9
31.92	72.02	750	625	77	180	8 x 75	8 x 75	439.42
L_7	L_8	C_{10}	L_9	C_{11}	C_{12}	C_{13}		
377.56	366.74	250	318.7	124.36	78.19	78.14		

With the targeted output power of 0.5 W at both 28 and 38 GHz bands, the power stage was determined to be designed using device with a gate periphery 1200 μm. To maintain the drivability of the driver stage, a 2-to-1 ratio of the transistors was adopted for the driver stage amplifier.

For dual-band designs, one of the critical issues relates to the roll-off of the MAG with respect to the different frequencies. This leads to substantial gain deviation between the 28 GHz and the 38 GHz band. Since the frequency response of the MAG is relating to the stability factor (k-factor), mitigation of the MAG roll-off could be possible by tweaking the profile of the k-factor with frequency. For a power amplifier design, an unconditionally stable condition must be satisfied for proper operation. The stability of the transistor also has a relation with the k-factor for a bilateral case with $S_{12} \neq 0$, which can be expressed as:

$$MAG = \Delta \cdot \frac{|S_{21}|}{|S_{12}|} \tag{2.17}$$

$$\Delta = k - \sqrt{k^2 - 1}, \ k \geq 1 \tag{2.18}$$

As shown in equation (2.17), the roll-off of MAG with frequency is depending on the two factors. Fig. 2.56 shows the two-cell stacked-FET configuration for further investigation using Keysight ADS simulator, where the gate periphery of the devices was set to be 4 × 75 μm. The discussion starts with the evaluation of the grounding capacitance effect on mitigation of the MAG profile of the stacked-FET.

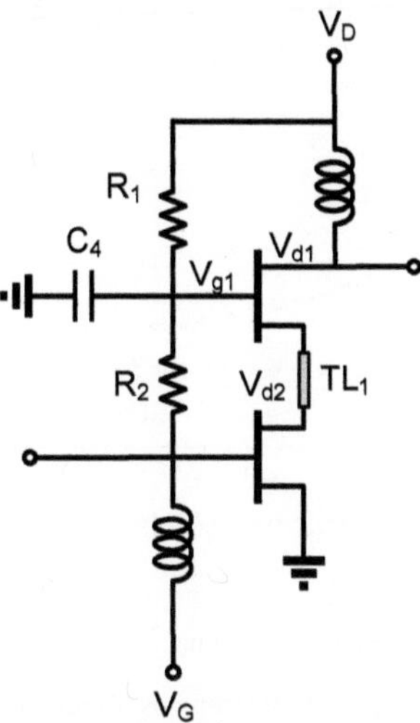

Figure 2.56. Cut out schematic of the 2-cell stacked-FET with resistive voltage divider from Fig. 2.55.

Fig. 2.57 and Fig. 2.58 show the frequency dependence of the two terms in (2.17) for the cases of the single device (8 × 75 μm) and the stacked-FET (2 × 4 × 75 μm), respectively.

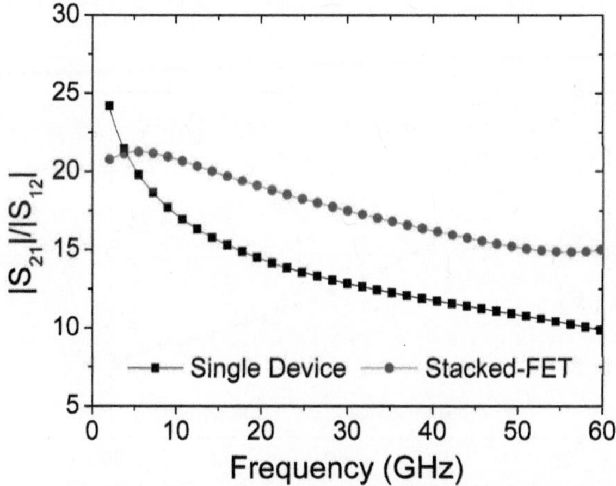

Figure 2.57. The frequency dependence plot of equation 2.17 for the single device and stacked-FET one.

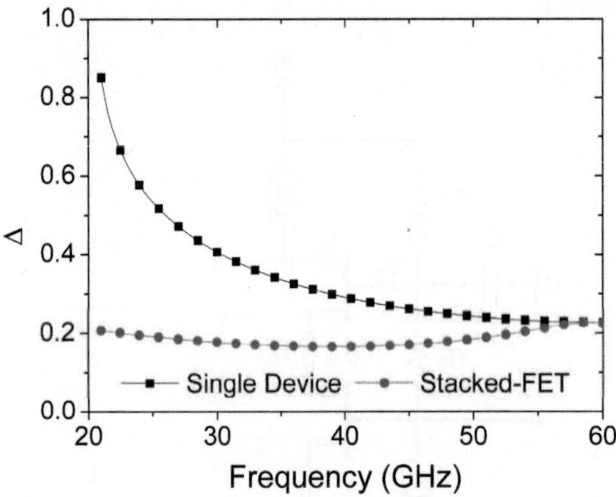

Figure 2.58. The frequency dependence plot of equation 2.18 for the single device and stacked-FET one.

For both cases, the $|S_{21}|/|S_{12}|$ degrades in the same trend as the frequency increases as shown in Fig 2.57. However, Fig. 2.58 shows that the stacked-FET configuration has a flatter response in the frequency of interest. These simulation results can help to evidence that the stacked-FET configuration can truly help smoothing the profile of the parameters for the MAG. As defined in equation (2.18), only when $k \geq 1$ satisfied the equation. Therefore, the frequency was plotted from 20 to 60 GHz as the k-factor of single device exceed 1 above 20 GHz. Next, the MAG as a function of frequency is plotted as Fig. 2.59. Obviously, the MAG roll-off is effectively mitigated by the stacked-FET configuration with the flatter response compared with single device.

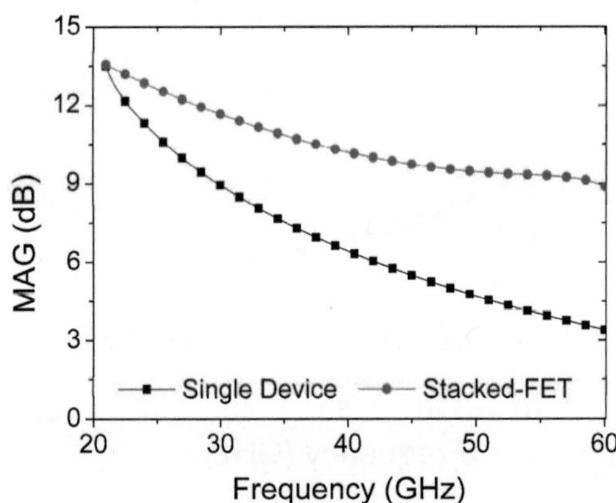

Figure 2.59. The frequency responses of MAG for the single device and the 2-cell stacked-FET configuration.

The grounding capacitance C_4 at the gate node of the common-gate transistor will strongly affect the performance of the stacked amplifier. First, the effect on voltage swings across the devices is simulated. Fig. 2.60 plots the voltage swing across the three nodes of the 2-cell stacked-FET.

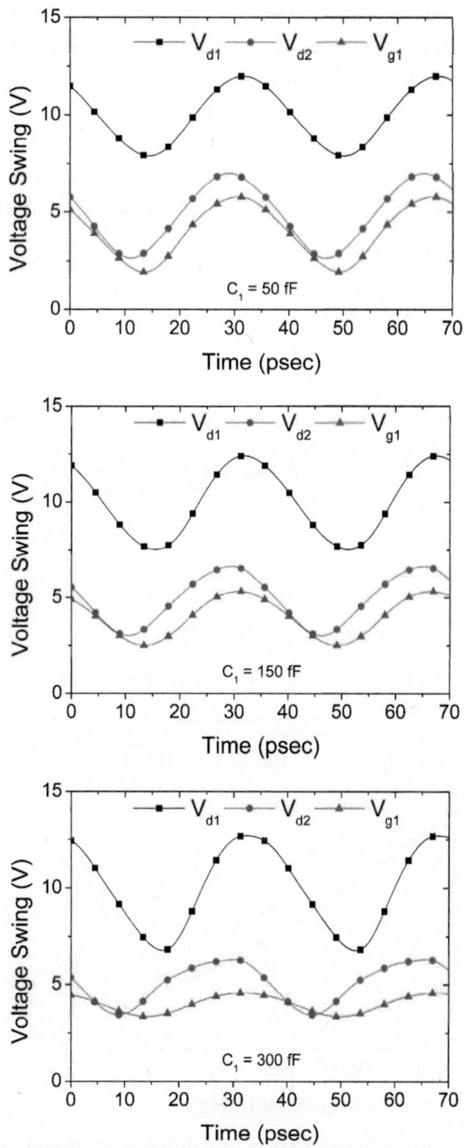

Figure 2.60. Voltage swings with different grounding capacitance.

From Fig. 2.60, an increase in the maximum voltage swing can be observed when the grounding capacitance increased. However, as the capacitance approaches 300 fF, a distortion at the peak region is observed. This may lead to oscillation while operating at high output level. The distortion at the distributed voltage swings may have led to the early drain-to-gate breakdown at the common-gate transistor on top. This condition will place a high risk for damaging the top transistor of the stacked amplifier during operation or measurement. For preventing potential oscillation at high output level, trade-off between the enhancement in voltage swings and reduction of distortion at peak level shall be made while selecting the grounding capacitance. Next, the effect of grounding capacitance on MAG and Δ will be discussed. Fig. 2.61 and Fig. 2.62 plots the frequency dependence of Δ and MAG with different grounding capacitance. The range of capacitance is limited to 50 fF to 150 fF based on the simulation results from Fig. 2.60.

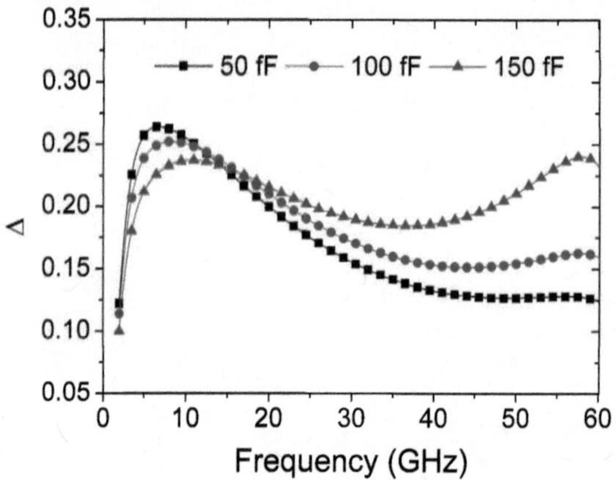

Figure 2.61. The frequency dependence of Δ with different grounding capacitance.

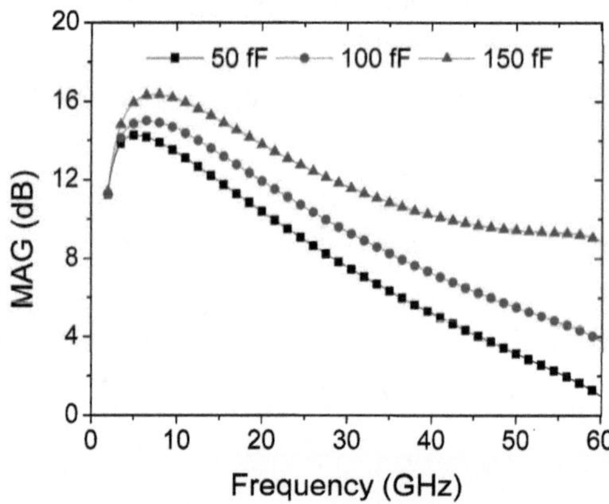

Figure 2.62. The frequency dependence of MAG with different grounding capacitance.

The capacitance value of 300 fF was not considered anymore as it was already determined to be dangerous with respect to a potential oscillation at high power operation. The figures shown above demonstrate that the magnitude of the grounding capacitance can tune the MAG as a function of frequency. For the profile of Δ, a flat curve is found at the Ka-band frequencies of interest. The proper selection of the grounding capacitance also led to a higher Δ for enhancement for the MAG. This effect can also be determined in Fig. 2.62 with the plotted MAG curve as functions of frequency. Obviously, an optimum choice of the grounding capacitance will strongly reduce the sensitivity of MAG to frequency. For the specific case of driver stage design, the capacitance was chosen to be 150 fF to effectively mitigate the deviation of MAG between 28 and 38 GHz. This has led to a gain deviation of 1.6 dB, which is better than the single device case of 2.8 dB.

Another design issue with the stacked-FET configuration is the effect of grounding capacitance on the optimum load impedance for maximum output power. Fig. 2.63 plots the power contours at 28/38 GHz and optimum impedance of the stacked-FET with different grounding capacitance at 5G NR FR2 frequencies. Note that the red contours are for 28 GHz and blue contours for 38 GHz.

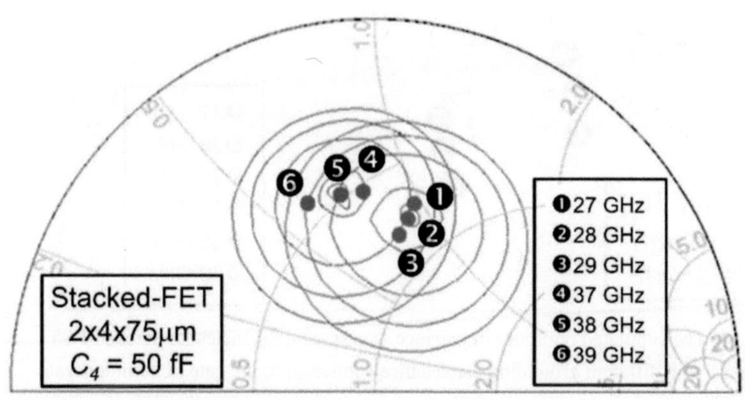

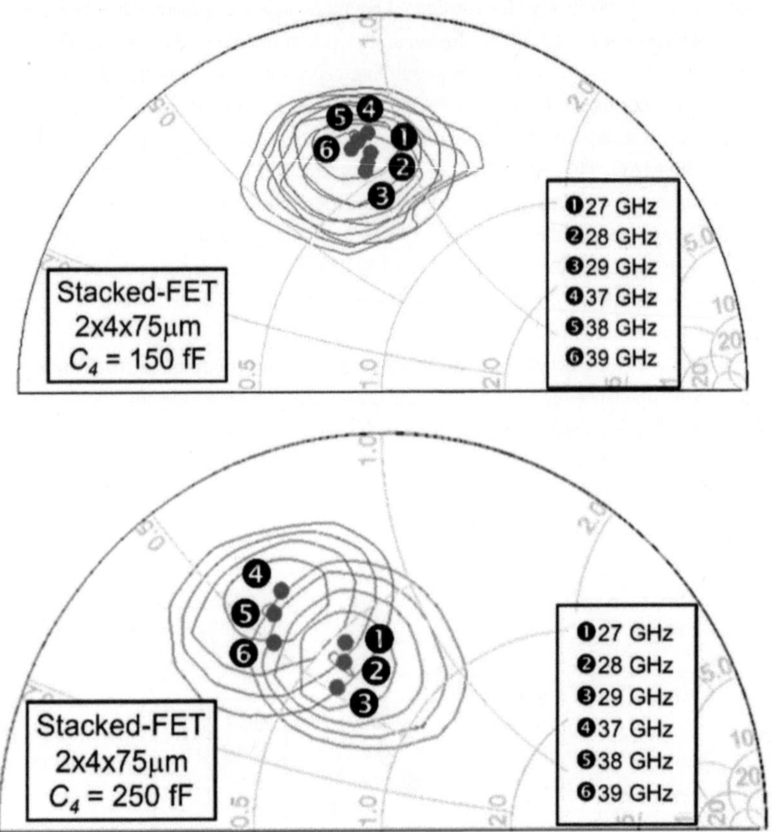

Figure 2.63. The simulated optimum impedance at various frequencies for stacked-FET (2×4×75 µm) with different grounding capacitance connected to the gate of common-gate device.

Obviously, the grounding capacitance will affect the power contours and the variation of optimum impedance between different operating frequencies. The selection of the grounding capacitance can be different based on the design specification and concern of the overall stability during operation. In this design approach, the major concern is attributed to mitigate the MAG profile and to maintain the stability during operation at high input level. Therefore, the grounding capacitance was chosen to be 150 fF based on the desired condition. For fairness of comparison, the simulated optimum impedance at various frequencies for single device is plotted in Fig. 2.64.

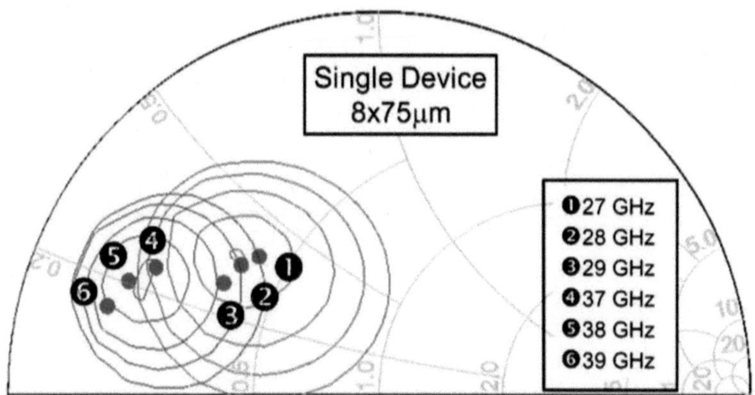

Figure 2.63. The simulated optimum impedance at various frequencies for single device (8×75 μm).

The optimum impedance simulated from 8×75 μm single device is showing a relatively large variation comparing with the stacked-FET case with a grounding capacitance of 150 fF. This was mainly due to the nature behavior of the transistor plotted in Fig. 1.6. The MAG normally follows the 20 dB/dec rule without specific structures applied to the transistor amplifier, which showed the advantage of the proposed stacked-FET approach.

A similar analysis was done for the power stage stacked-FET using two 8×75 μm devices, where a grounding capacitance of 250 fF was selected. The optimum impedance for maximum output power (Z_{opt2}) was simulated as (18 + j 23) Ω / (13.6 + j 23.8) Ω at 28/38 GHz. The corresponding source impedance (Z_{S2}) was (4.6 + j 0.3) Ω / (3.4 - j 1.2) Ω at 28/38 GHz. These terminations lead to a P_{sat} of 28.7/28.5 dBm with 9.5/9-dB small-signal gain at 28/38 GHz by load-pull simulation. The optimum impedance was matched to the 50 Ω load with a three-step LC matching network. Fig. 2.64 shows the impedance looking into the output matching network. The matching path with notation referring to the Smith Chart is plotted in Fig. 2.65 for 28 and 38 GHz.

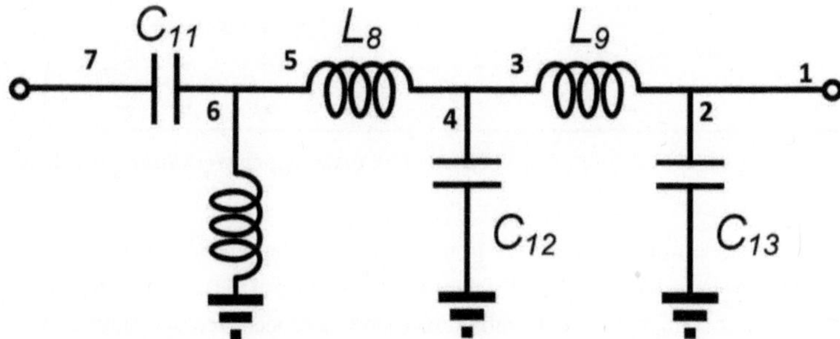

Figure 2.64. Output impedance looking into the matching network.

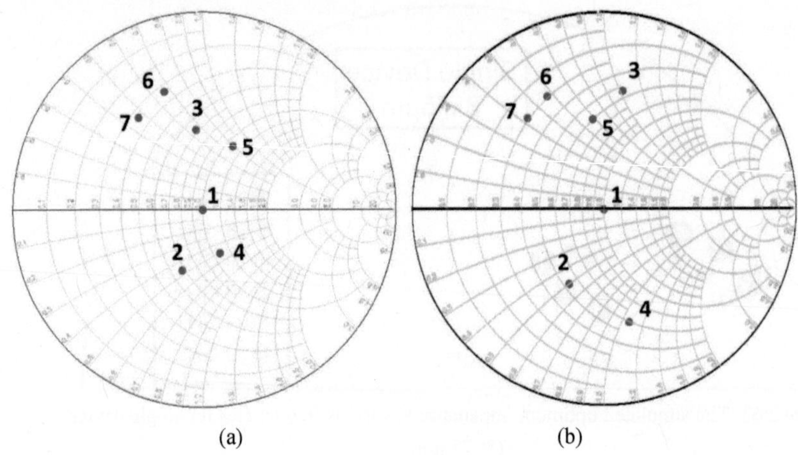

(a)   (b)

Figure 2.65. The matching path of output network for (a) 28 GHz and (b) 38 GHz.

The insertion loss induced by the output matching network was simulated to be 1.2/0.86 dB at 28/38 GHz. Such arrangement was to mitigate the 0.5-dB small-signal gain deviation between the two frequencies simulated using load-pull simulator. The impedance matching contour from 28 to 38 GHz is plotted in Fig. 2.66 with the power contours simulated from load-pull simulator.

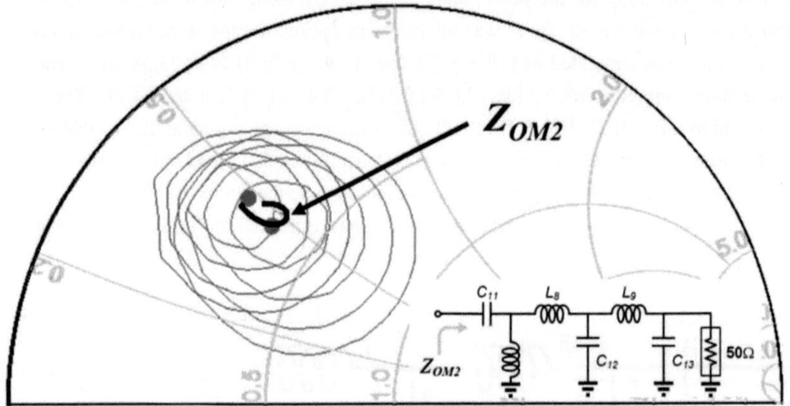

Figure 2.66. The impedance contour (Z_{OM2}) of the output matching network from 28 GHz to 38 GHz.

Although the matching network was designed to match the optimum impedance at 38 GHz, a nice power match was also achieved at 28 GHz due to the proximity of the optimum impedance benefited from the stack configuration with optimum grounding capacitance. The simulated P_{sat} (with output matching network) was 28.5/28.2 dBm at 28/38 GHz.

Similar design procedure has been applied for the interstage and input matching design. Fig. 2.67 shows the impedance looking the output of interstage matching network. The matching path with notation referring to the Smith Chart for the interstage is plotted in Fig. 2.68 for 28 and 38 GHz. With such arrangement, a matching loss of 1.02/0.76 dB was achieved at 28/38 GHz.

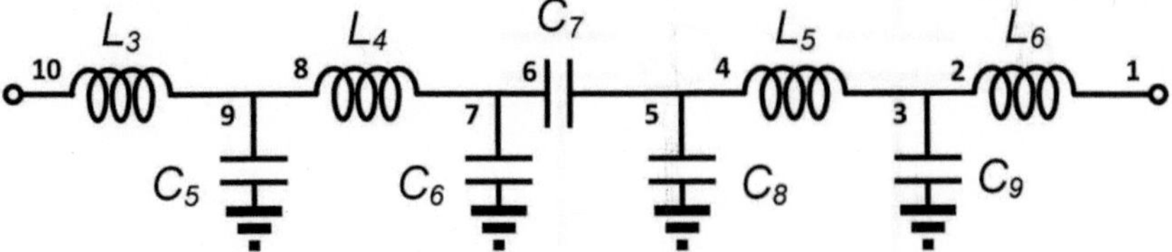

Figure 2.67. Impedance looking into the interstage matching network.

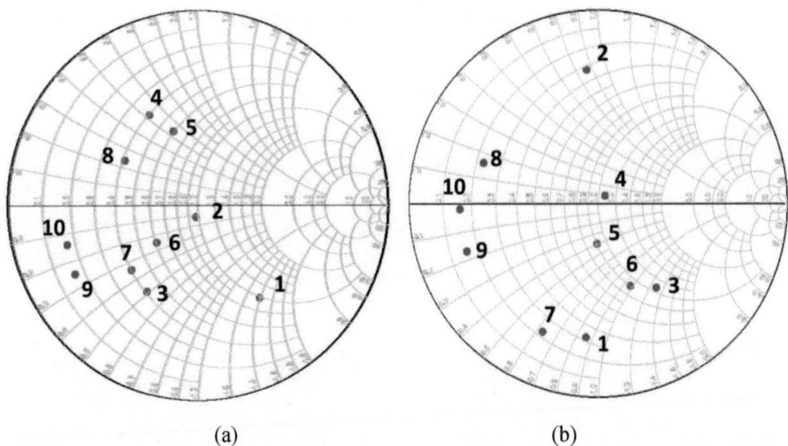

(a)   (b)

Figure 2.68. The matching path of interstage network for (a) 28 GHz and (b) 38 GHz.

Fig. 2.67 shows the impedance looking the output of interstage matching network with notation referring to the Smith Chart. The matching path for the input of driver stage is plotted in Fig. 2.70 for 28 and 38 GHz. With such an arrangement, a matching loss of 0.94/0.64 dB was achieved at 28/38 GHz.

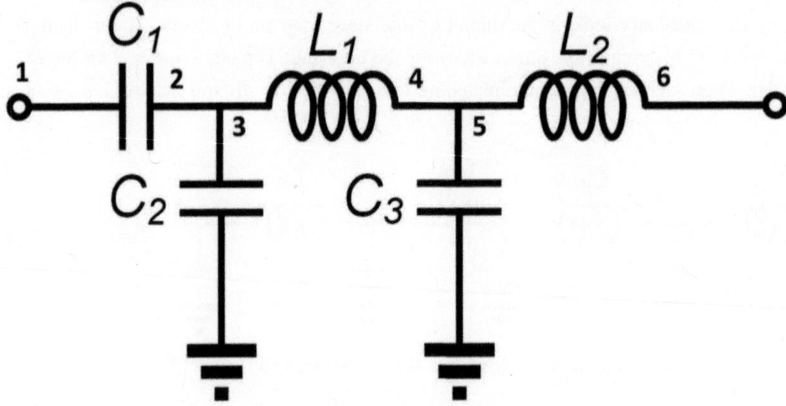

Figure 2.69. Impedance looking into the input matching network.

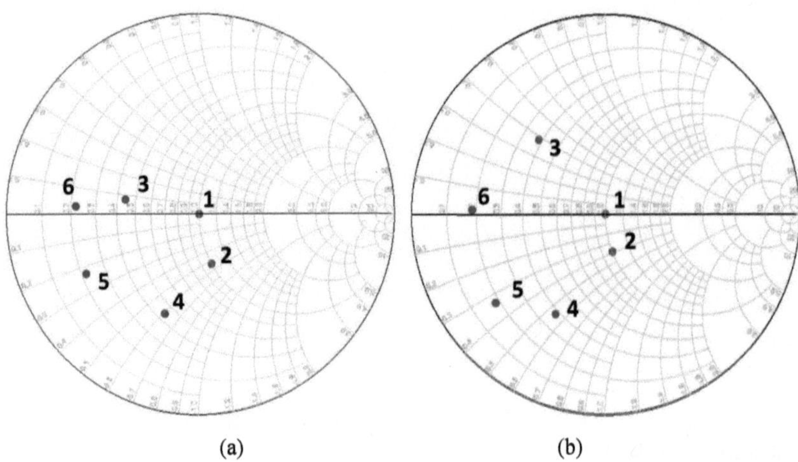

(a)   (b)

Figure 2.70. The matching path of interstage network for (a) 28 GHz and (b) 38 GHz.

Overall, the two-stage design achieved a small-signal gain of 17.8/17.4 dB, a P_{sat} of 28.5/28.2 dBm, and PAE of 41%/39% at 28/38 GHz. The simulated input/output return loss is greater than 11/9 dB at both frequency bands of interest.

2.4.3. Measurement results

Fig. 2.71 shows the chip photo of the two-stage dual-band power amplifier fabricated using 0.15-μm GaAs pHEMT technology from WIN Semiconductors. The chip size is 2.7 mm × 1.7 mm with all the probing pads for on-wafer measurements.

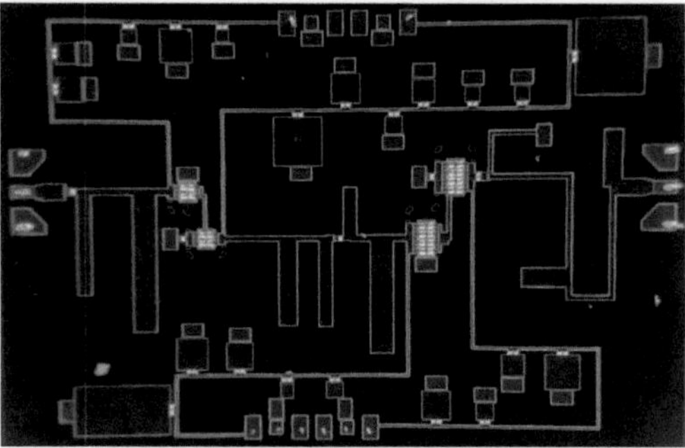

Figure 2.71. Chip photo of the fabricated dual-band power amplifier.

The measurement setup for small-signal characterization is plotted in Fig. 2.72. Fig. 2.73 shows the measured results from 20-45 GHz, where the bias voltages were set as V_{D1} = 10 V, V_{G1} = -1.0 V, V_{D2} = 10 V and V_{G2} = -1.0 V for class-AB biasing condition. The corresponding currents of the driver and power stages were 38 mA and 80 mA, respectively. The simulated response of the two-stage dual-band power amplifier is also included for comparison. The measured input and return loss exceeds 10 dB at both bands. The dual-band power amplifier delivered a small-signal gain of 18.5/18 dB at 28/38 GHz. Fig. 2.74 plots the measured stability k- and B1 factor, which shows an unconditionally stable across the frequency of interest. The measurement setup for small-signal characterization is plotted in Fig. 2.75. Fig. 2.76 and 2.77 shows the measured gain and PAE versus output power at 28 and 38 GHz under CW mode excitation. As shown, the PA delivered a P_{sat} of 28.5/28.3 dBm and the peak PAE was measured to be 39%/36% at 28/38 GHz, showing a less difference in both small- and large-signal performance between the two frequency bands.

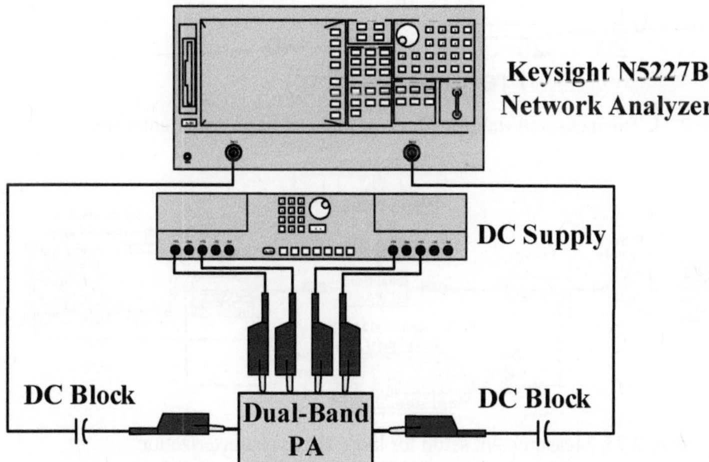

Figure 2.72. Measurement setup for small-signal characterization.

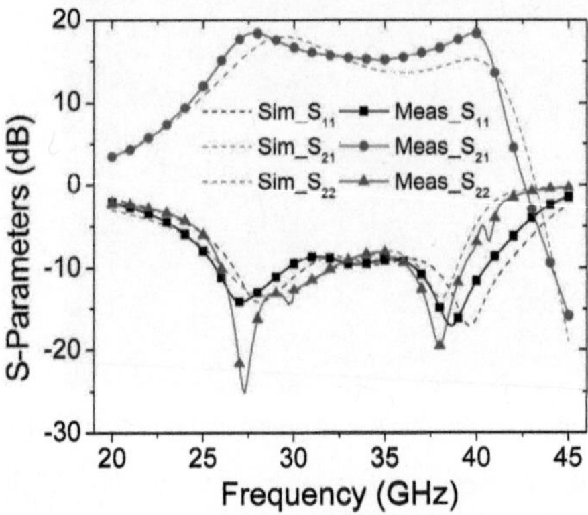

Figure 2.73. The measured and simulated S-parameters of the dual-band power amplifier.

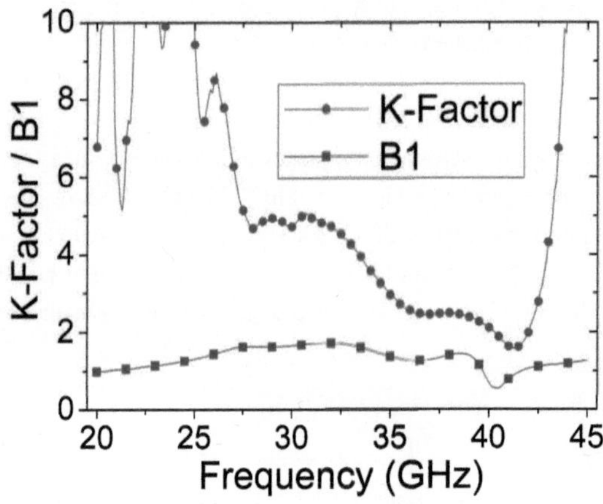

Figure 2.74. The measured stability factor of the dual-band power amplifier.

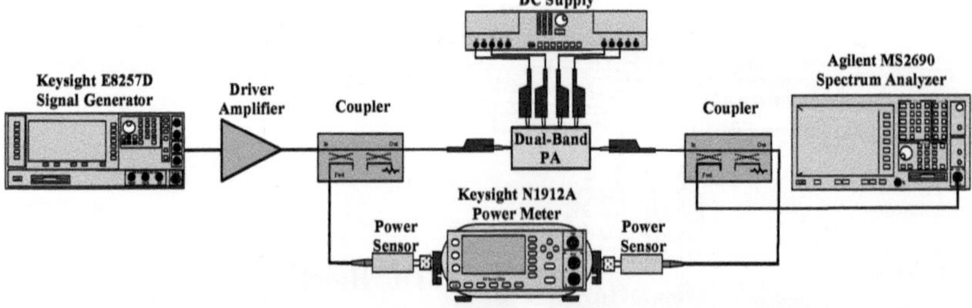

Figure 2.75. Measurement setup for large-signal characterization.

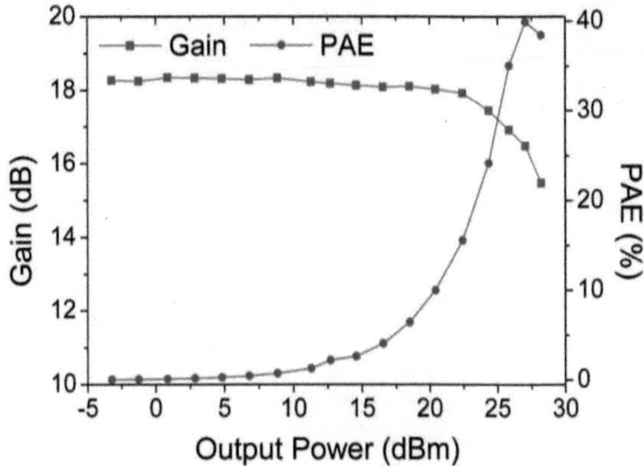

Figure 2.76. The measured power performance of the dual-band power amplifier at 28 GHz.

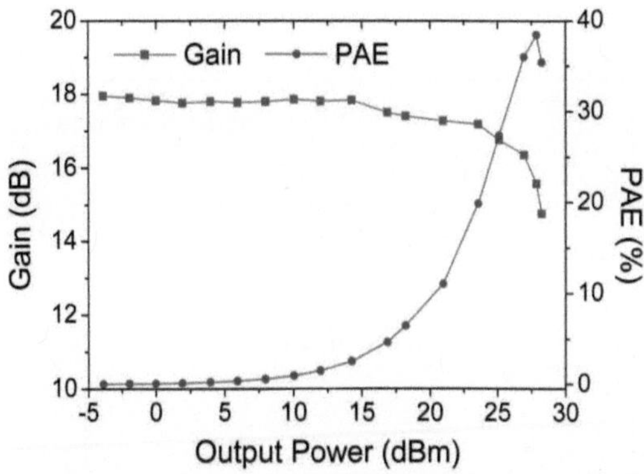

Figure 2.77. The measured power performance of the dual-band power amplifier at 38 GHz.

The measured P_{sat} and PAE as functions of frequencies are plotted in Fig. 2.78, exhibiting almost identical performance at both bands of interest.

The dual-band power amplifier was also characterized using modulated signals. Fig. 2.79 shows the setup for modulation measurement. The 64-QAM with pseudorandom binary sequence 11 (PRBS11) signal and 0.35 roll-off factor (α) of the square-root-raised-cosine pulse-shaping filter (SRRC) were used to test the dual-band PA. Fig. 2.80 shows the measured EVM of 64-QAM with 1.5 Gbps data rate versus different average output power at 28/38 GHz. The measured EVM was under 4.8% at the average output power level of 23.3 dBm at 28 GHz. At 38 GHz, the EVM was under 4.7% at the average output power level of 22.8 dBm. Fig. 2.81 shows the measured constellation and the corresponding spectrum of 64-QAM with 1.5 Gbps data rate at the average output power of 23.3/22.8 dBm at 28/38 GHz.

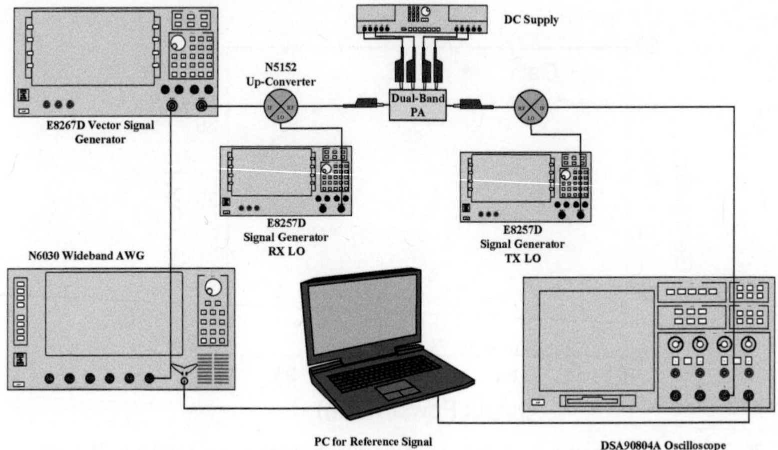

Figure 2.79. Measurement setup for modulation measurement.

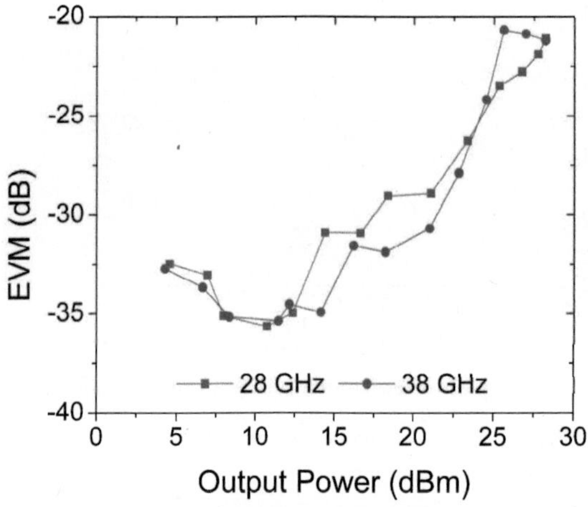

Figure 2.80. The measured EVM versus average output power at both bands.

To check the reliability of the fabricated PA under different ambient temperatures, an on-wafer S-parameter characterization was performed. In Fig. 2.82, the measured results of the small-signal gain at different temperature levels. The temperature coefficient of the small-signal gain is extracted to be -0.0039 dB/°C, exhibiting a relatively good thermal stability.

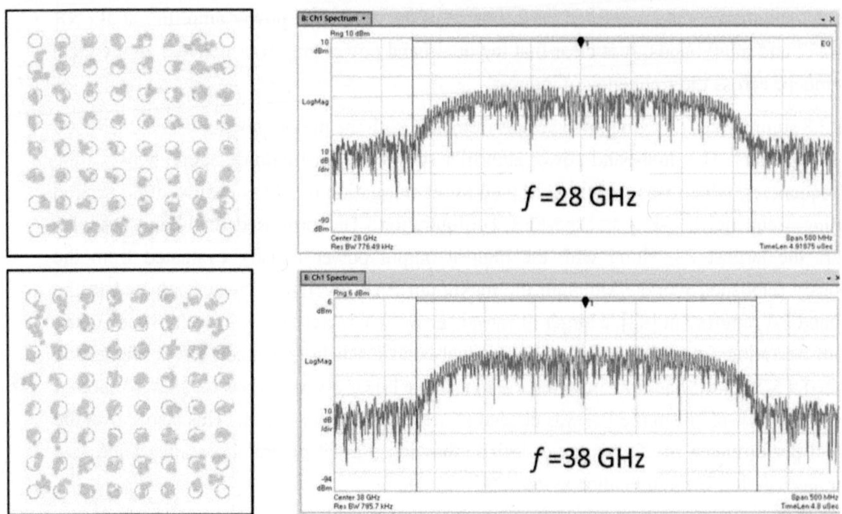

Figure 2.81. Measured constellation, EVM and spectrum of 64-QAM with 1.5 Gbps data rate at the average output power of 23.3/22.8 dBm at 28/38 GHz.

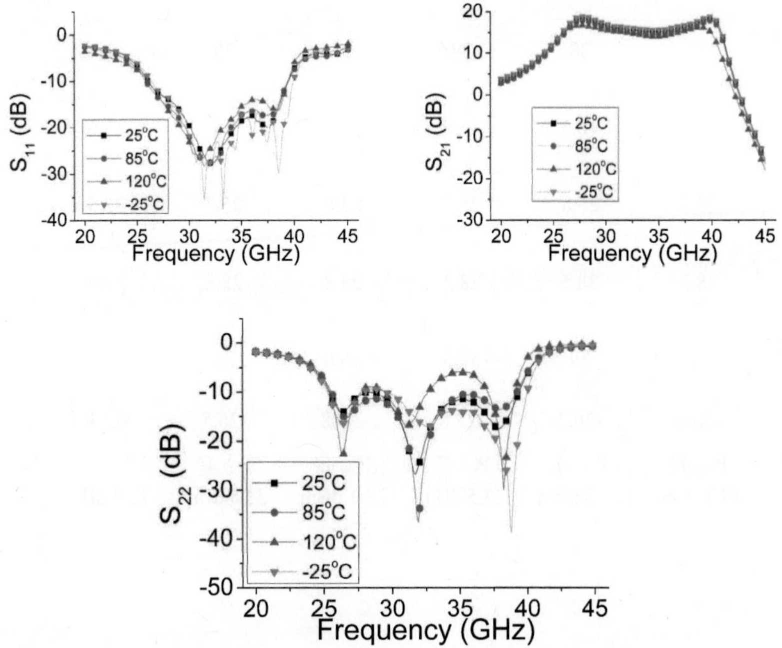

Figure 2.82. The measured small-signal performance under different ambient temperatures.

Table 2.7 summarizes the measured performance of the dual-band power amplifier at 5G NR FR2 n257 and n260 bands. It is clear that the dual-band power amplifier delivers an identical performance across the frequency of interest.

Table 2.8 summarizes the comparison of performance with the previously published dual-band power amplifiers. The dual-band power amplifier shows the least gain deviation between the two frequencies, which was mainly attributed to the mitigation of MAG profile at the desired operating frequencies using stacked-FET configuration. The proposed dual-band PA also delivers the largest P_{sat} along with the highest output power density evaluated from the previously published works. Although the maximum PAE level was lower compared to the dual-band PA reported in [83], a larger linear power was characterized using modulated signal. At both operating frequencies, the proposed dual-band PA using stacked-FET configuration exhibited a 5.2/5.4 dB back-off from the P_{sat} to achieve linear operation. As for [83], a reported 7.7/10.7 dB back-off from the P_{sat} was needed to achieve its linear region. This also shows the identical performance of the proposed design, showing a great potential for integration in a carrier aggregation communication system.

Table 2.7 Measured performance of the dual-band power amplifier at 5G NR FR2 bands.

Band	n257			n258		
Frequency (GHz)	27	28	29	37	38	39
Gain (dB)	18	18.5	17.8	17.4	18	17.6
P_{1dB} (dBm)	25.7	25.8	25.7	25.6	25.7	25.5
P_{sat} (dBm)	28.3	28.5	28.2	28.2	28.3	27.9
Peak PAE (%)	37	39	37.3	36.6	36	35.4
EVM (dB)	-26.9 P_{out} @ 23.1 dBm	-26.2 P_{out} @ 23.3 dBm	-25.8 P_{out} @ 23.5 dBm	-26.8 P_{out} @ 23.1 dBm	-26.5 P_{out} @ 22.8 dBm	-25.9 P_{out} @ 22.4 dBm

Table 2.8 Performance comparison with previously published works.

Reference	Dual-band PA		[81]		[82]		[83]		[84]	
Technology	0.15-µm GaAs		0.18-µm SiGe		0.13-µm SiGe		90-nm GaAs		0.15-µm GaAs	
Topology	Stacked-FET		Concurrent filtering		Multiband DPA		Concurrent		Optimal matching contour	
Freq. (GHz)	28	38	25.5	37	28	37	29.6	39	28	39
Gain (dB)	18.5	18	21.4	17	18.2	17.1	19.1	20.3	20	12
ΔG^+ (dB)	0.5		4.4		1.1		1.2		8	
P_{1dB} (dBm)	25.8	25.7	10.4	7.1	15.2	15.5	N/A		20	12
P_{sat} (dBm)	28.5	28.2	16	13	16.8	17.1	23.8	23.7	21.9	22.7
P_{sat}/Area (mW/mm²)	154.2	144	45.2	22.7	27.2	29.1	137.9	134.7	103.3	124.2
Peak PAE (%)	39	36	10.6	4.9	20.3	22.6	51.3	49.5	28.9	32.6
Signal Type	1.5 Gbps 64-QAM		N/A		3 Gbps 64-QAM		6 Gbps 64-QAM		N/A	
EVM (dB)	-26.2	-26.5	N/A		-27	-30.3	-32	-30.8	N/A	
P_{Lin} (dBm)	23.3	22.8	N/A		9.2	9.5	16.1	13	N/A	

+Difference in gain between the two frequencies.

3. Antenna design and module integration

As the applications at millimeter-wave and beyond requires a larger equivalent isotropically radiated power (EIRP), phased-array antenna systems are considered as a solution with the improvement of transmitter gain and output power with multiple antenna elements and RF front-end modules included. However, the phased-array antenna requires a careful design as the antenna elements might have interference with each other, such as the sidelobes. Therefore, the location of the antenna elements must be arranged based on the radiation profile. In the next section, a review of the antenna configurations for millimeter-wave applications will be discussed.

3.1. Antenna configurations

Based on the requirements of 5G wireless communication link, high gain antenna approaches are necessary to compensate the high free space loss. For antenna targeting for point-to-multipoint (PtM) and point-to-point (PtP) communications, the key design parameters for each of the cases are plotted in Fig. 3.1 and 3.2. As shown in the figures, each of them requires different radiation characteristics as the operation mode differs.

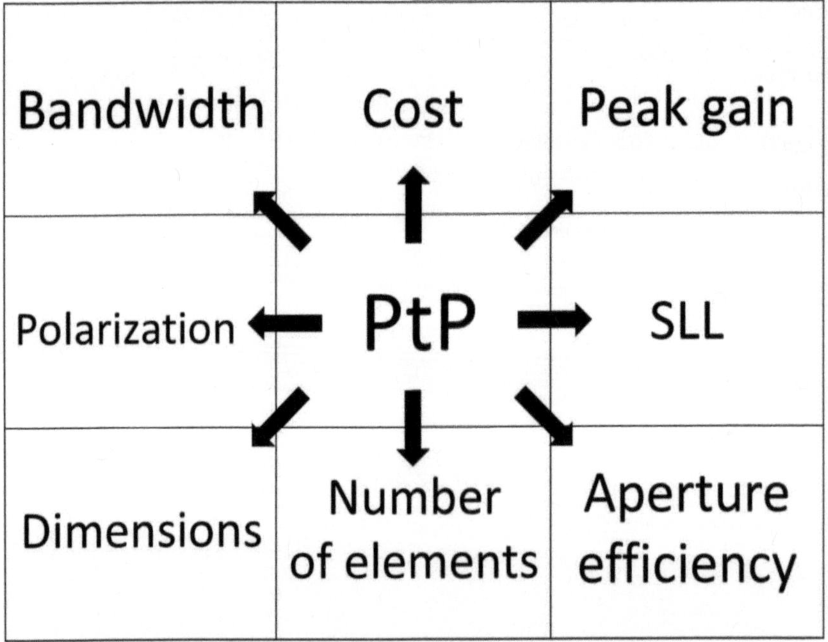

Figure 3.1. Key design parameters of antenna for point-to-point communication.

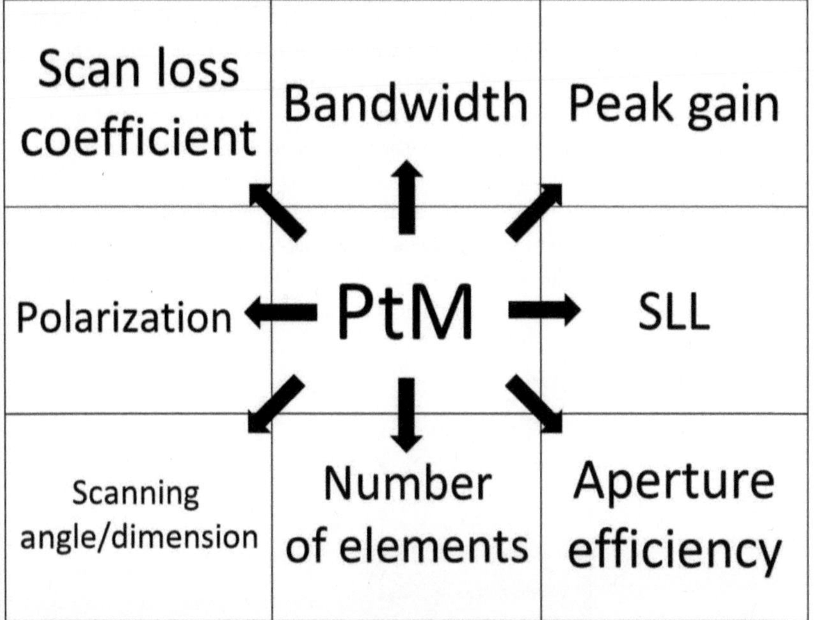

Figure 3.2. Key design parameters of antenna for point-to-multipoint communication.

For instance, as PtP has a firm direction for signal transmission, the dimension matters; when it comes to the PtM application, the scanning angle and dimension will need to considered as this will determine the available operating range for the array system.

As different antenna configurations are showing different advantages and disadvantages, an overview of the common structures applied for millimeter-wave applications will be discussed.

First, the microstrip patch antenna is commonly reported as the manufacturing procedure is relatively simple. A conventional patch antenna is shown in Fig. 3.3.

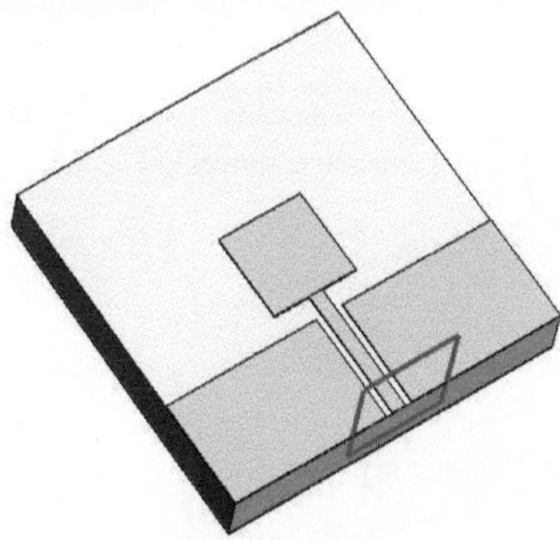

Figure 3.3. 3D view of a conventional patch antenna.

The patch antenna configuration is often applied for point-to-multipoint communication as it offers a radiation pattern with wide coverage. This characteristic makes a wide scanning angle and dimension possible at array level, which makes it a competitive candidate for point-to-multipoint applications. Another advantage for patch antenna is the simple control of the polarization, as it can achieve both linear and circular polarization. For system integration at millimeter-wave frequencies, a compact size of the antenna element is required for further scaling of the phased-array system. However, as the dimension of the patch antenna is proportional to operating wavelength, antenna size at millimeter-wave frequencies is usually very small leading to possible reduction in bandwidth. In order to overcome the bandwidth reduction induced by the size reduction, a shorting loaded patch antenna was reported in [86]. The proposed configuration is shown in Fig. 3.4.

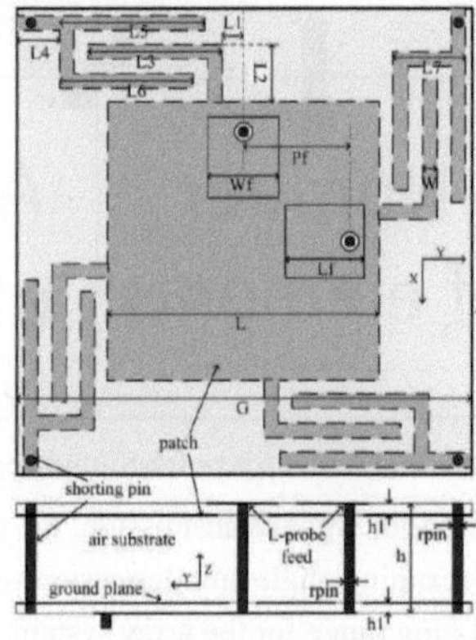

Figure 3.4. Structure of the shorting loaded patch antenna [86].

The patch antenna also utilized a coupled shorting strip to compensate the parasitic inductance to equivalent a capacitor for LC resonator. With the loaded coupled shorting strip, the compact circularly polarized patch antenna achieved a compact dimension of $\lambda/7 \times \lambda/7 \times \lambda/15$ and an improvement of 7.2% in terms of the operating bandwidth comparing with the conventional patch antenna approach.

As targeting for broadband operation, the Vivaldi antenna is considered as a common solution. The traditional configuration for a Vivaldi antenna is shown in Fig. 3.5.

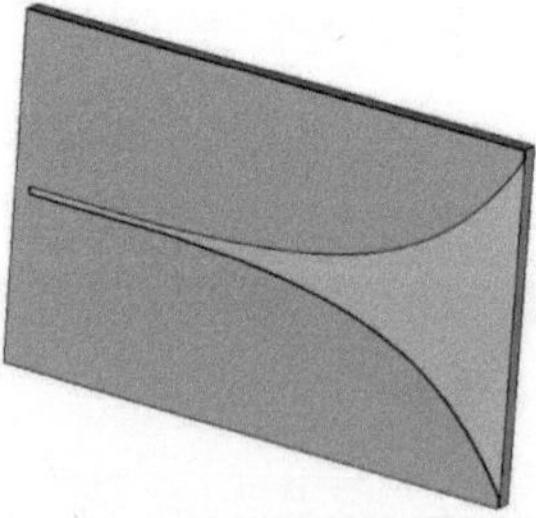

Figure 3.5. Traditional configuration of a Vivaldi antenna [87].

Such antenna configuration is normally fabricated using dielectric substrate, which makes it feasible for integration with external components. Other advantages with the Vivaldi antenna are the high gain characteristic and directional radiation. As shown in Fig. 3.5, the Vivaldi antenna has a fan-shaped structure connecting to the 50 Ω feeding line as an impedance matching at the bottom layer. The front-to-back ratio of the Vivaldi antenna is considered as a key parameter for gain enhancement. In [88], slot structures were attached to the Vivaldi antenna for perturbation of the surface current of the radiation patch, which have led to a high directivity. Such arrangement helped to increase the front-to-back ratio by 6.5 dB, resulting in an increment of the maximum gain of the antenna by 1.5 dBi. Fig. 3.6 shows the structure of the reconfigurable Vivaldi antenna reported in [88].

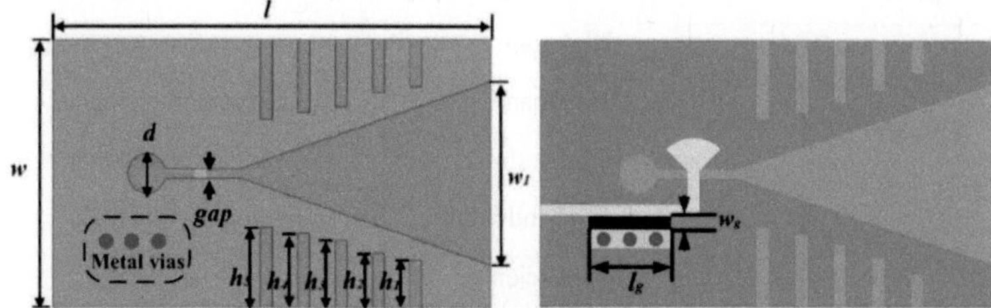

Figure 3.6. Structure of the reconfigurable Vivaldi antenna with optimized slots [88].

With the implementation of tunable resistor with graphene nanoplate, a dynamic manipulation of the antenna gain was achieved with a controllable resistance. A graphene-based nanoplate was introduced to provide different resistance while switching the supplied voltage. A tuning range from 30 to 210 ohm was demonstrated with the graphene-based nanoplate, which helped to create different beam directions. This antenna is fed from the backside of the substrate using a microstrip line, then the signal flows from backside up to the frontside by the metal vias for further signal transmission to the air.

Open-ended waveguide antennas were reported to have high gain and high radiation efficiency, which make them popular for millimeter-wave applications. However, such configuration is not easy for integration with chips and a bulky size makes it less attractive for commercial applications. To solve this challenge, a substrate-integrated waveguide (SIW) was developed. SIW technology is a compelling candidate for designing cost-effective, high-gain, low-loss systems in millimeter-wave technologies.

Table 3.1 summarizes the pros and cons of the popular antenna configurations at millimeter-wave frequency for phased-array system.

Conventionally, waveguide-based horn antennas were dominant for communication systems at millimeter-wave frequencies. Such designs were initially selected because of the good radiation performance in terms of gain and directivity. However, the waveguide type antenna has the major drawback of not being well suited for integration with planar circuits. A counterpart approach, the substrate integrated waveguide (SIW), was invented as the replacement of the metallic waveguide solution in which the metallic walls were implemented by "via fences" and the substrate dielectric served as the filling material of the waveguide. Such a configuration has the advantage of an easy implementation in standard layered printed-circuit-board process but additional waveguide-to-microstrip transitions are required which may limit the operating bandwidth. In this dissertation, the dual-exponential tapered slot antenna (DETSA) configuration have been adopted. This planar type antenna shows an as nice performance as the patch antenna and is easy for system level integration will chip mounting and further packaging purpose. Figure 3.8 shows the conceptual schematic of the antenna with design parameters included.

Table 3.1. Pros and cons of different antenna configurations.

Configuration	Advantage	Disadvantage
Vivaldi	High gain Broadband Symmetric radiation pattern Tunable bandwidth	Large dimension Complex fabrication
SIW	High efficiency Radiation pattern High power handling Low conductor loss	Narrow band Leakage losses Low operating frequency
Patch	Integration Low cost Dual polarization Multi-band operation	Low gain High cross polarization Low power handling Low impedance bandwidth

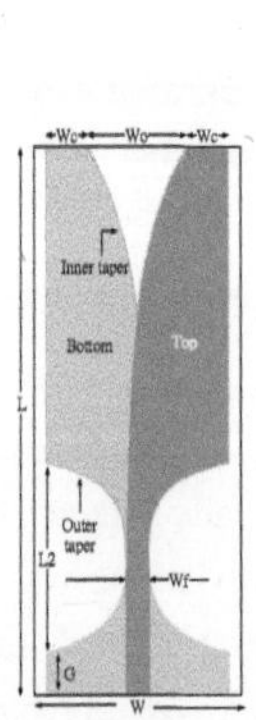
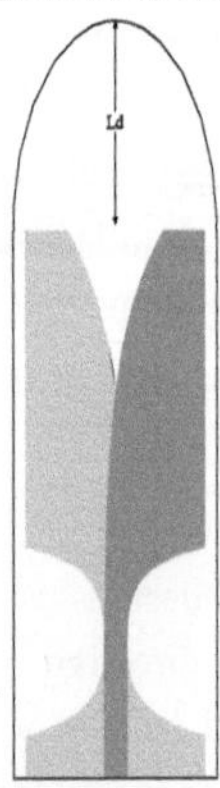

Figure 3.8. The conceptual schematic of the proposed DETSA antenna.

3.1.1. Dual exponential taper slot antenna (DETSA) design and simulation

The DETSA antenna radiates in the end-fire direction and the concept of elliptic dielectric loading can be applied for gain and directivity enhancement along the beam direction. In such design, a balun transformer is also included at the transition region making it extremely easy for integration with planar circuits since the feed of the antenna can share the ground plane with the MMIC chip.

For all the antennas discussed in this section, the conductors are located in the x-y plane. Such arrangement forces the H-plane to the y-z plane ($\varphi = 90°$) and E-plane to the x-y plane ($\theta = 90°$). The design will be based on the AlN substrate material. The electrical properties of the AlN substrate is summarized in Table 3.2. The simulation of the DETSA antenna was done using CST Studio Suite [89].

Table 3.2. Electrical properties of the AlN substrate.

Property	Units	
Density	g/cm^3	≥ 3.25
Surface roughness Ra	µm	≤ 0.5 (As-fired)
		≤ 0.5 (Lapped)
		≤ 0.05 (Polished)
Thermal conductivity	W/mK	≥ 170
Thermal expansion	$\times 10^{-6} K^{-1}$	4.6
Breakdown voltage	kV/mm	20
Dielectric constant	@1 MHz	9.0
Dielectric loss tangent	$\times 10^{-4}$ @1 MHz	≤ 10
Thickness	mm	0.25

The simulation setup for the DETSA antenna using AlN substrate in CST Studio Suite is shown in Fig. 3.9. For this DETSA antenna approach, an antipodal feeding structure was included for the design, which allows for a balanced feed for the antenna by applying a microstrip line feeding structure.

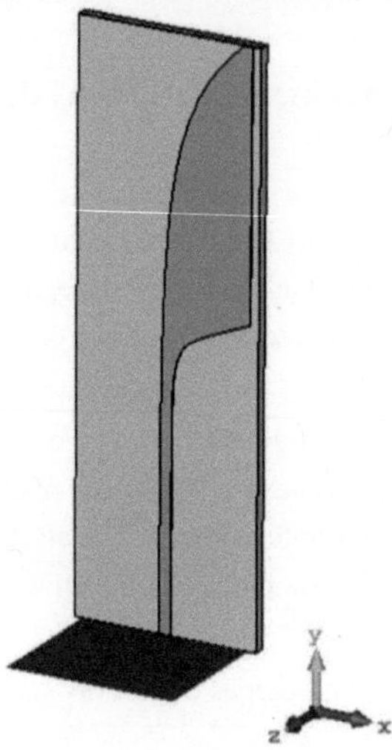

Figure 3.9. Simulation setup for DETSA antenna.

For the design of the DETSA antenna, the outer and inner taper of the structure can be described as:

$$x = \pm c_s \exp\left(k_s y^{rf}\right) \tag{3.1}$$

$$x = \pm c_w \exp\left(k_w y^{sf}\right) \tag{3.2}$$

Note that x represents the distance from the centered line of the slot to the inner and outer edge of one conductor. The design parameters of the DETSA antenna is summarized in Table 3.3.

Table 3.3. Design parameters of the DETSA antenna.

Parameters	Dimension (mm)
W_c	0.2
W_o	3.25
W_f	0.25
L_2	4.5
G	2
W	4.5
L	12

The simulated input return loss is plotted in Fig. 3.10. With a nice impedance match from 33 GHz up to 50 GHz, the DETSA antenna is well feasible for application at the 5G NR FR2 n260 band (37 – 40 GHz). To further investigate the radiation pattern of the DETSA antenna, simulation of the farfield at 38 GHz is shown in Fig. 3.11. A peak gain of 4.94 dBi is observed from the simulated radiation pattern, where the direction of the main lobe is located at 14° from the center with a 3-dB angular bandwidth of 99.9°. The side lobe level was simulated to be -0.9 dB, respectively.

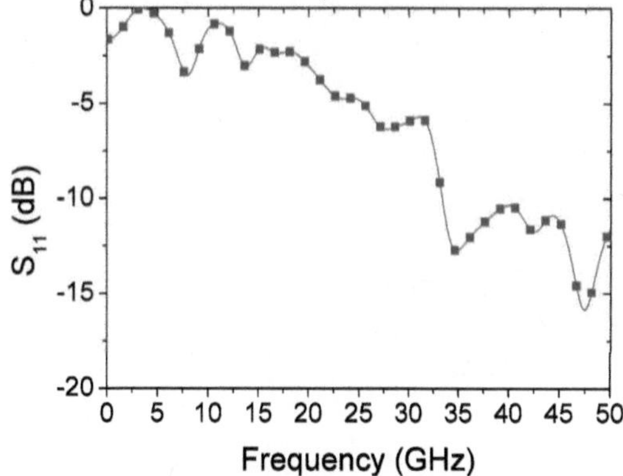

Figure 3.10. Simulated S_{11} for the DETSA antenna.

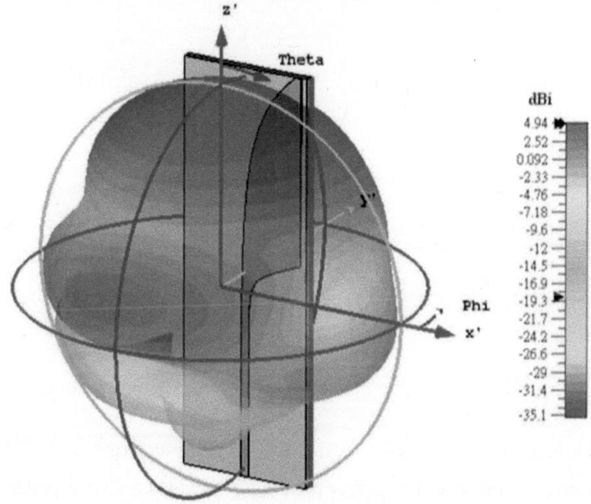

Figure 3.11. Simulated radiation pattern in 3D view of the DETSA antenna.

Next, the design of the DETSA antenna array will be performed. Fig. 3.12 shows the setup of 1 x 4 DETSA antenna array. The feeding network was designed using Wilkinson power dividers for proper distribution of the excitation signal. The simulated input return loss of the DETSA antenna array is plotted in Fig. 3.13.

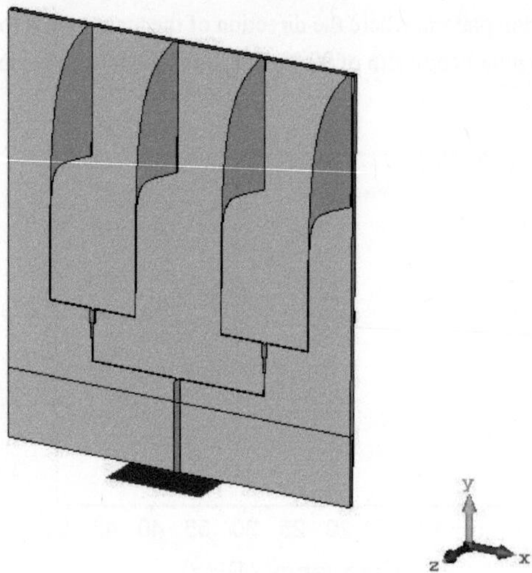

Figure 3.12. Simulation setup for DETSA antenna array.

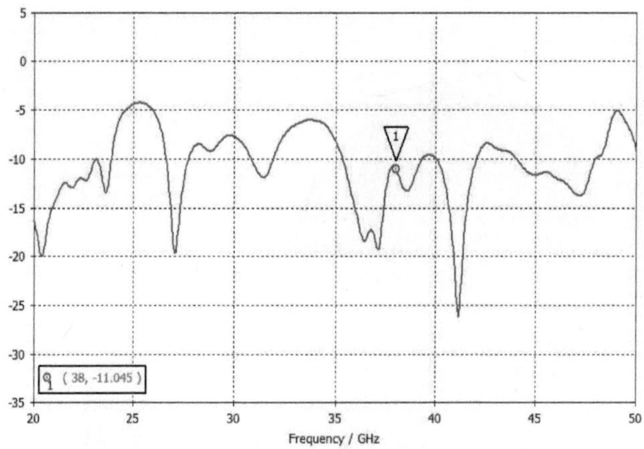

Figure 3.13. Simulated S_{11} for the DETSA antenna array.

Although the Wilkinson power divider limits the operating bandwidth, a nice impedance match from 35 GHz up to 41 GHz has been obtained. Thus, the DETSA antenna array is proven feasible for application at 5G NR FR2 n260 band (37 – 40 GHz). Fig. 3.14 shows the radiation pattern of the DETSA antenna array at 38 GHz. A peak gain of 8.33 dBi is observed in main beam direction.

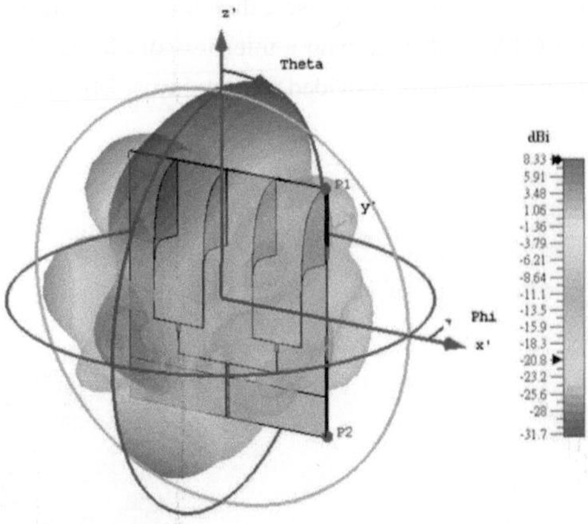

Figure 3.14. Simulated radiation pattern in 3D view of the DETSA antenna array.

3.1.2. Technological implementation and measurement results

The DETSA antenna element and array is fabricated using FBH in-house process on polycrystalline AlN substrate material. The fabricated DETSA antenna and array is shown in Fig. 3.15.

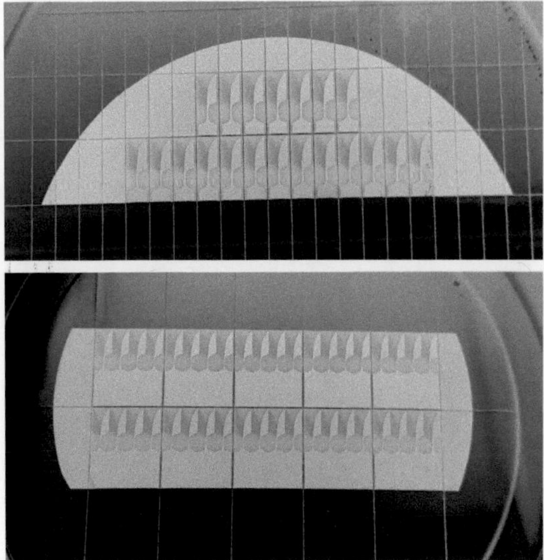

Figure 3.15. Fabricated DETSA antenna and array using FBH in-house AlN process.

For the measurements connectorized setups were used, therefore the antenna element and the array were mounted on the RO4003 substrate with a thickness of 20 mil. The connectorized DETSA antenna with endlaunch connector provided by Southwest Microwave are shown in Fig. 3.16.

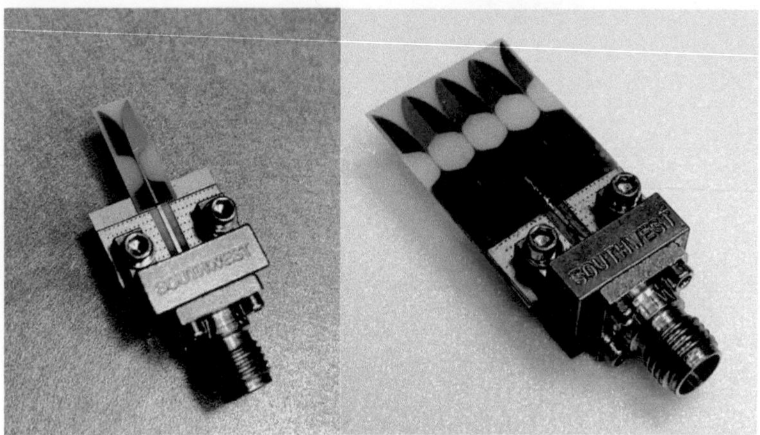

Figure 3.16. Connectorized DETSA antenna and array for characterization.

The feeding line of the mounted substrate is connected with the AlN DETSA antenna and array using wirebonding technique. A close view of the wirebonding connection to the antennas is shown in Fig. 3.17.

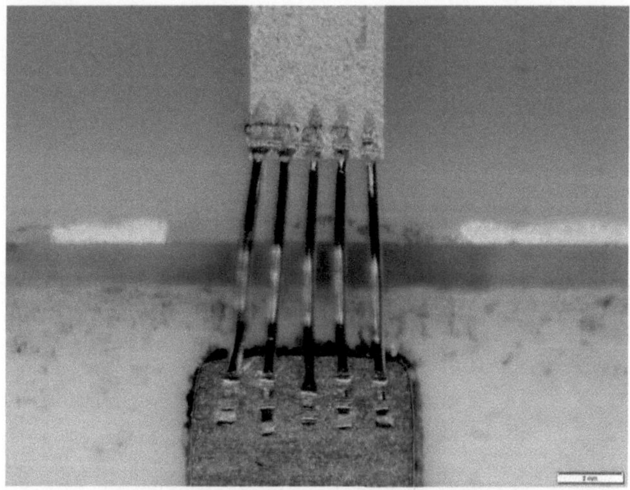

Figure 3.17. View of wirebonding connection.

The small-signal characterization was done using Keysight N5227B PNA network analyzer. Fig. 3.18 and Fig. 3.19 plots the measured input return loss of the DETSA antenna and array from 0.1 GHz to 45 GHz. Both of them show a nice impedance match covering the targeted operating frequency of 37 – 40 GHz, where the input return loss is less than 7 dB.

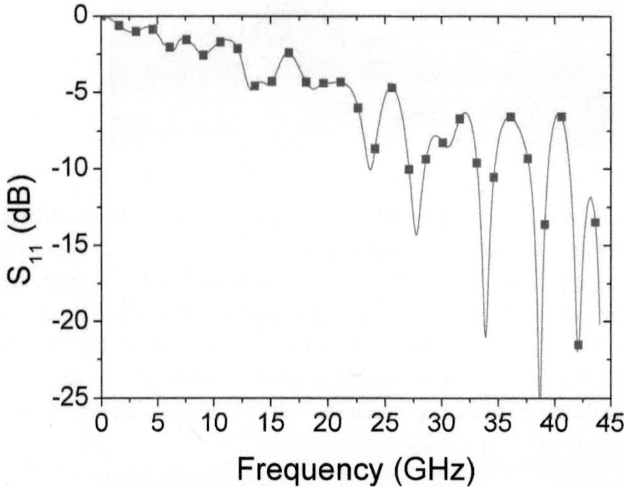

Figure 3.18. Measured input return loss of the DETSA antenna.

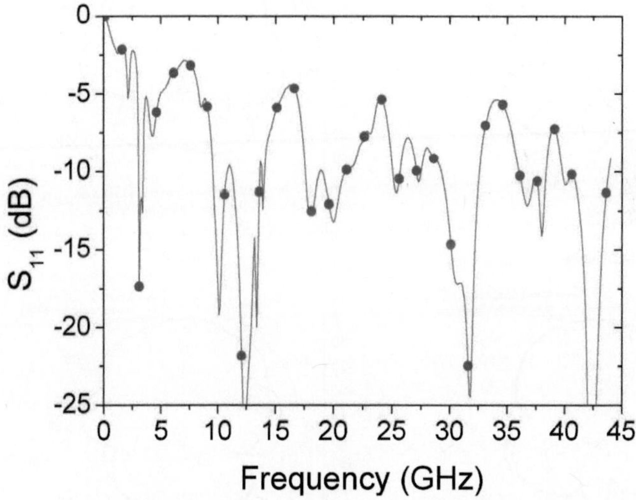

Figure 3.19. Measured input return loss of the DETSA antenna array.

The radiation patterns of the antenna arrangements where then characterized in an anti-echoic chamber according to Fig. 3.20.

Figure 3.20. Measured input return loss of the DETSA antenna array.

The measured radiation pattern of the DETSA antenna at 38 and 39 GHz with simulation results included are plotted in Fig. 3.21. The measured peak gain is plotted in Fig. 3.22, where a gain of 5 dB is characterized at 38 GHz, respectively. The radiation pattern of the DETSA antenna array at 38 and 39 GHz with simulation results included are plotted in Fig. 3.23. The measured peak gain is plotted in Fig. 3.24, where a gain of 8.44 dB has been obtained at 39 GHz. The simulated pattern nicely matches with the measurement results for both, DETSA antenna and array. The deviation of the antenna pattern between 120 and 150° is due to the specific set-up in the anti-echoic chamber.

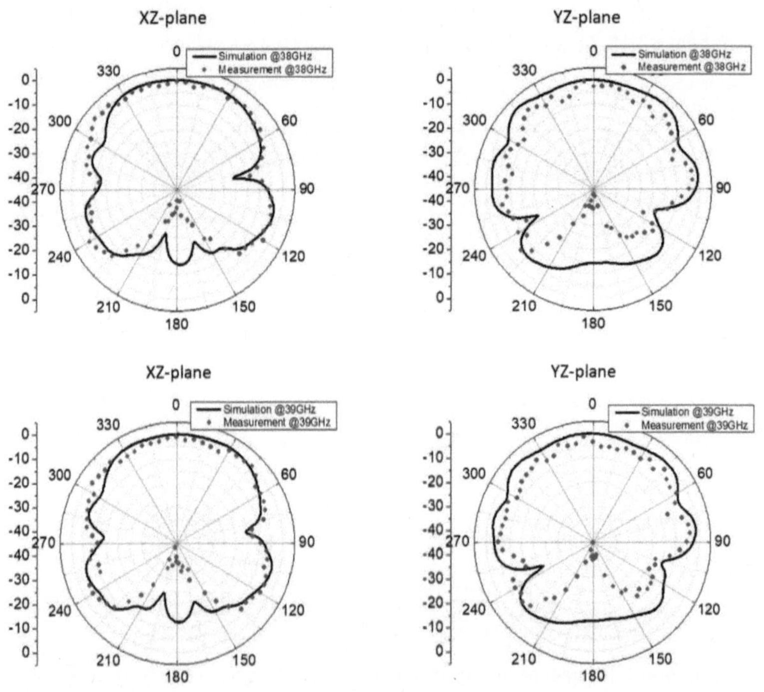

Figure 3.21. Measured radiation pattern of the DETSA antenna at 38 and 39 GHz.

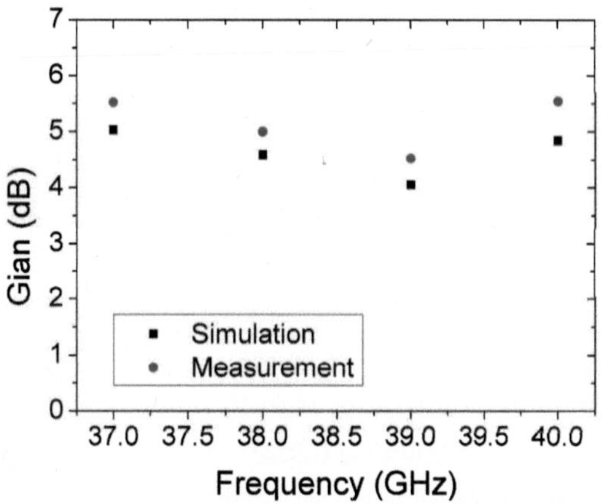

Figure 3.22. Measured peak gain of the DETSA antenna.

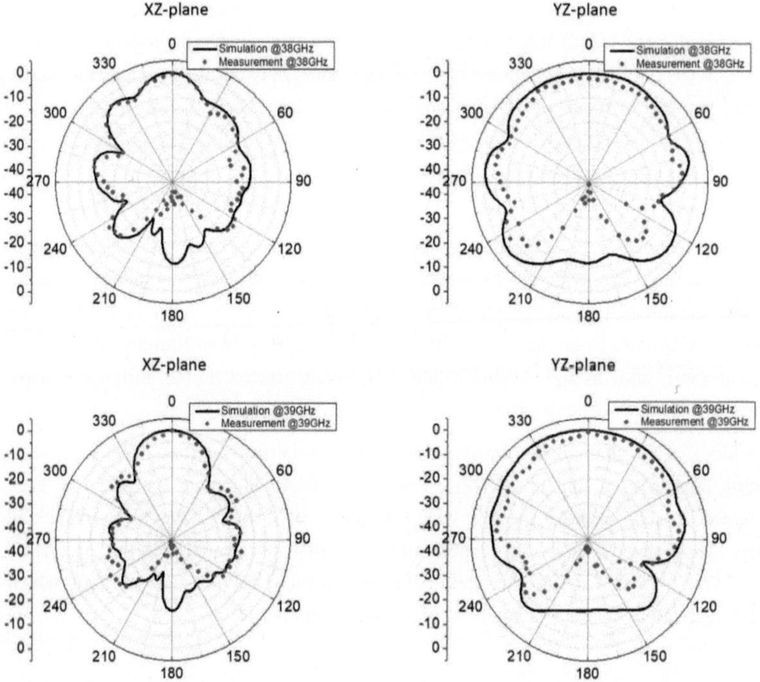

Figure 3.23. Measured radiation pattern of the DETSA antenna array at 38 and 39 GHz.

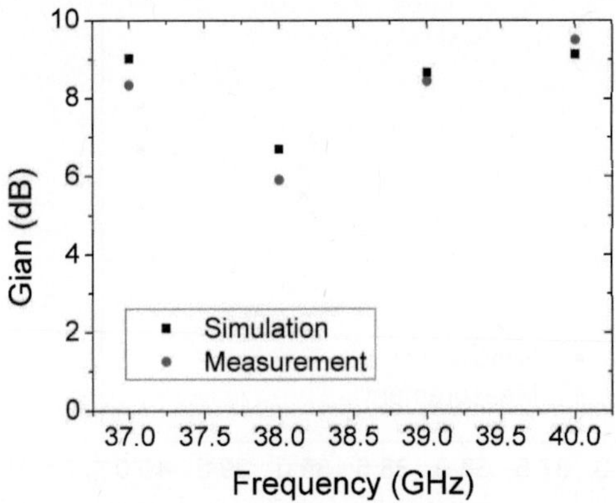

Figure 3.24. Measured peak gain of the DETSA antenna array.

The measured gain of the array antenna didn't achieve an ideal 6-dB improvement compared to the single antenna. The possible reason may be the high dielectric constant of the substrate. Such characteristic has led to the potential surface wave coupling reducing the radiation efficiency. Moreover, the tilted radiation pattern on the feeding network is also one of the reason degrading the gain performance of the array antenna.

3.2. Beam steering network configurations

For applications at millimeter-wave frequencies, the high free space path loss for long-range transmission is a huge challenge for engineers. Besides of the phased-array antenna system, beamforming technique was often applied to enhance the directivity and adaptivity of the overall system. Common beam steering technique such as analog beamforming [89], digital beamforming [90] and hybrid beamforming [91] were reported for millimeter-wave applications.

A building block of the analog beamforming antenna system is shown in Fig. 3.25. For analog beamforming approach, analog phase shifters are designed to modulate the phase of each antenna element to achieve the purpose of beamforming. One of the disadvantages of such an arrangement is associated with the degradation of the antenna characteristics at high operating frequencies. This is mainly due to the phase difference induced by the phase shifters, which makes it complicated to enhance the gain of the overall antenna system.

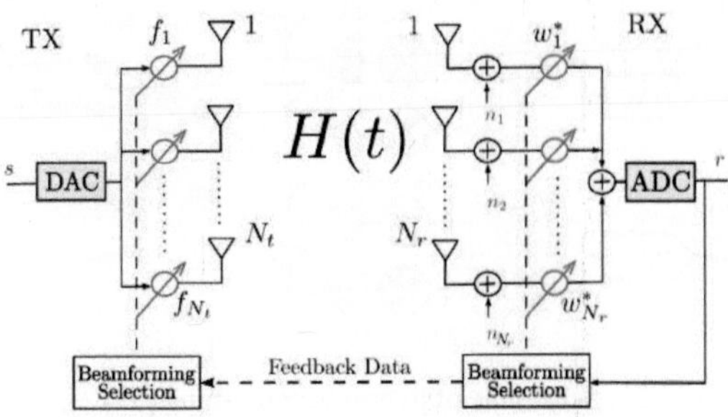

Figure 3.25. Schematic of analog beamforming [89].

As for the digital beamforming system, the schematic is shown in Fig. 3.26. Digital beamforming technique forms the beam by using digital circuits operating at baseband. The operation starts with the conversion of RF signals to baseband signals, which helps to improve the adaptive characteristic of the antenna array. Such arrangement helps to enhance the efficiency of beamforming. However, such arrangement stringently enlarges the complexity of the overall system, which makes further system integration a huge task to deal with.

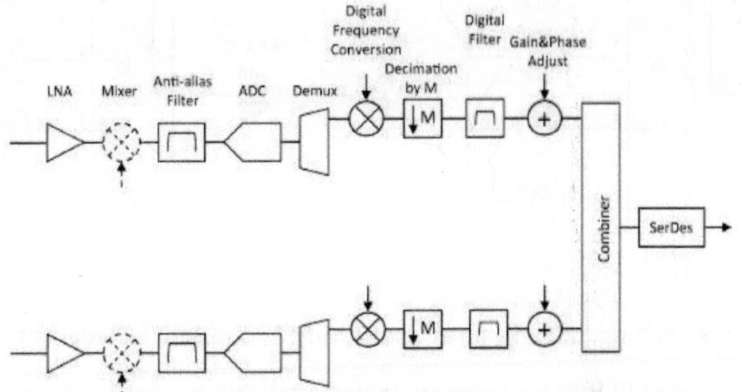

Figure 3.26. Schematic of digital beamforming [90].

To combine the advantages for analog and digital beamforming techniques, a hybrid beamforming approach was adopted. Such arrangement combines the operating mode for analog and digital beamforming. For instance, the phase difference between antenna elements are set by using analog phase shifters and the phase error will then be fixed by the baseband digital beamforming technique to reduce the number of RF components. The accuracy of beamforming can then be improved. It is important to mention that the complexity of the overall system can be reduced compared to the conventional digital beamforming. The schematic of the hybrid beamforming technique is shown in Fig. 3.27.

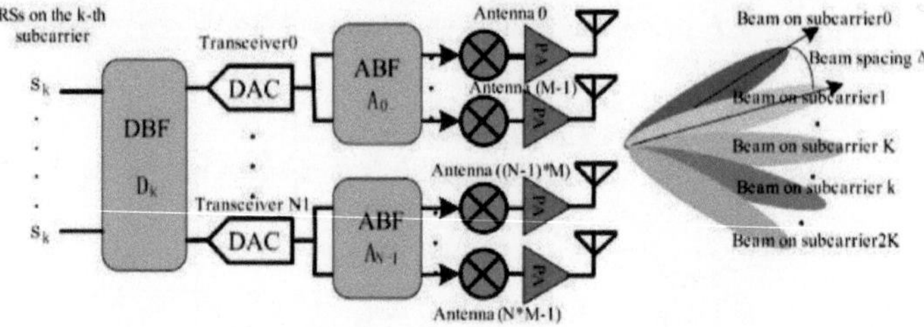

Figure 3.27. Schematic of hybrid beamforming [91].

Beam switching approach is also a common method for beam steering at millimeter-wave frequencies. The building block for beam switch system is plotted in Fig. 3.28. As the insertion loss of phase shifters operating at Ka-band frequencies and beyond are relatively large compared to frequencies below 20 GHz [92]-[94], other design approaches for phase shifting were reported to overcome this issue.

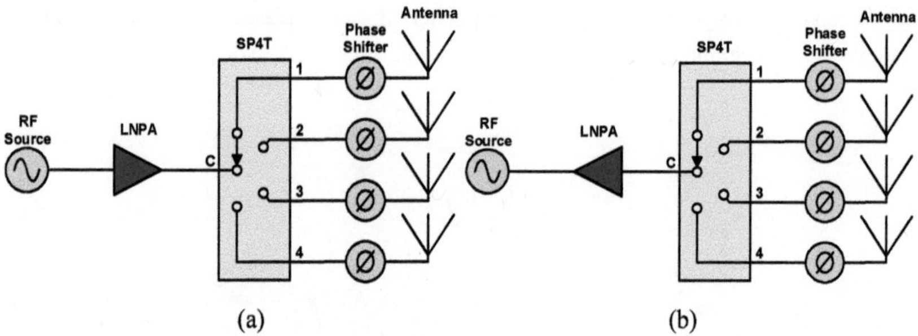

Figure 3.28. Beam switching antenna at (a) transmitting mode and (b) receiving mode.

A Rotman Lens was reported in [95] for beam switching application. The simple configuration has an advantage in large operating bandwidth and low fabrication cost. The reported 60 GHz phased-array antenna using Rotman Lens is shown in Fig. 3.29. A high resistive silicon substrate was chosen for reducing the conversion loss induced by the transmission lines. A patch antenna array was designed for beam switching purpose and to also enhance the gain of the system. A radiation efficiency of 50-70% was reported with exciting different channels.

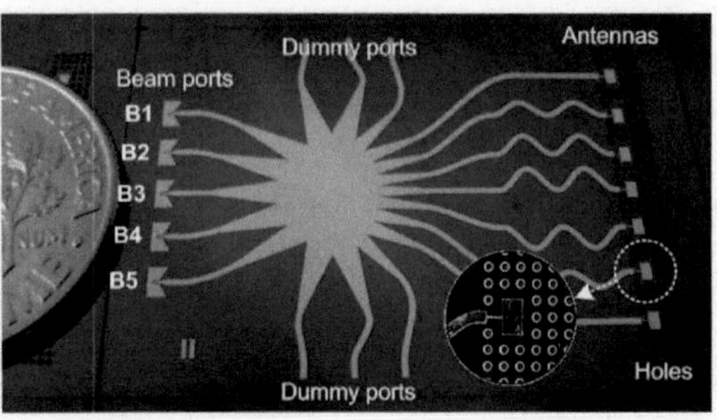

Figure 3.29. Photo of the phased-array antenna using Rotman Lens [96].

The Butler matrix is another common design approach for phased-array antenna system with beam switching functionality. The advantages of Butler matrix are the wide operating bandwidth and the well controllable angle for beam excitation. A beam switchable vertical off-center-fed dipole antenna array using butler matrix was presented in [96]. Such arrangement covered a wide operating frequency from 57 – 66 GHz with a maximum antenna gain of 15.6 dBi at 60 GHz.

In this section, a design of beam switchable antenna array using Butler matrix will be presented.

3.2.1. Butler matrix with Single-Pole-Four-Throw switch

Fig. 3.30 shows the conventional schematic of a 4 × 4 Butler matrix. It consists of 4 hybrid couplers, 2 cross-overs and 2 45° phase shifters using transmission lines.

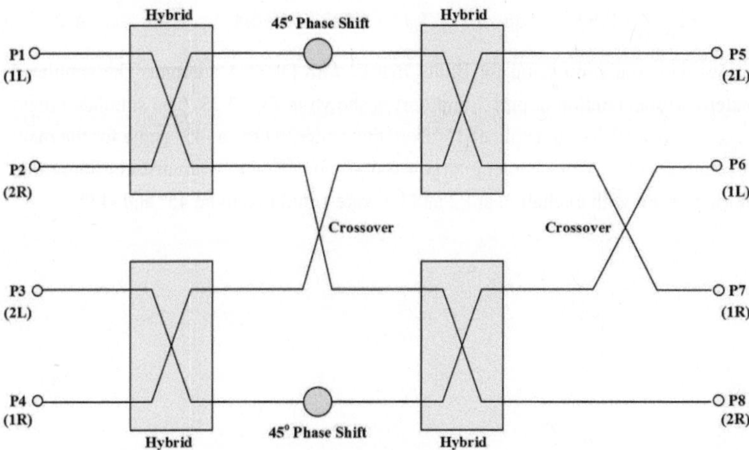

Figure 3.30. Schematic of Butler matrix.

The input signal is switched to the 4 terminals P1 to P4 and thus creates a systematic phase shift at the terminals 5 to 8. These terminals will then directly drive the different antenna pixels. For the following simulation, the Butler matrix will be designed using Rogers RO3010 substrate with a thickness of 0.25 mm, a dielectric constant of 10.2 and a dielectric loss tangent of 0.0022 @ 10 GHz. The simulation setup of the butler matrix using coplanar waveguide approach is shown in Fig. 3.31.

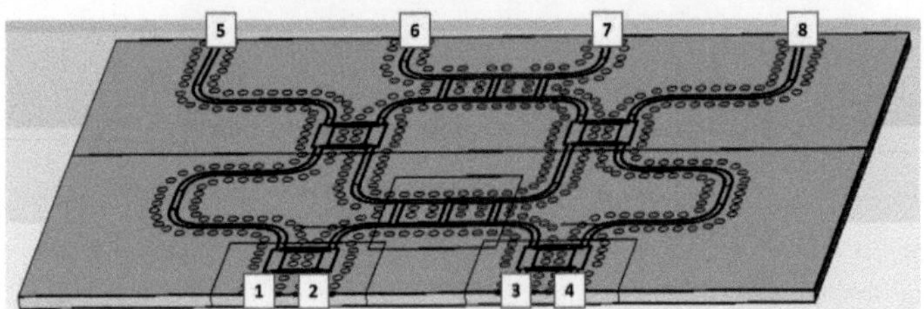

Figure 3.31. Simulation setup for Butler matrix using CST Studio.

The simulated results of the phase and insertion loss of the CPW Butler matrix is summarized in Table 3.4.

Table 3.4. Simulated performance of the Butler matrix.

Input \ Output	Port 1 (1L)		Port 2 (2R)		Port 3 (2L)		Port 4 (1R)	
	Phase (°)	IL (dB)	Phase (°)	IL (dB)	Phase (°)	IL (dB)	Phase (°)	IL (dB)
Port 5	45	8.7	136	8.5	89	7.2	179	5.3
Port 6	92	8.4	0	8	226	5.2	135	8.7
Port 7	134	8.1	223	4.8	3	9.6	89	8.0
Port 8	180	5.5	86	7	133	8.4	45	8.5

Fig. 3.32 shows the simulation setup for Butler matrix with DETSA antenna. The simulated radiation pattern with excitation at port 1 and port is shown in Fig. 3.33. The simulated main beam with excitation at P1 has an angle of 15° from the center, where a -45° angle for the main beam with excitation at P3. As the Butler matrix was designed using a symmetric configuration, the angle of main beam with excitation at P2 and P4 were simulated to be 45° and -15°.

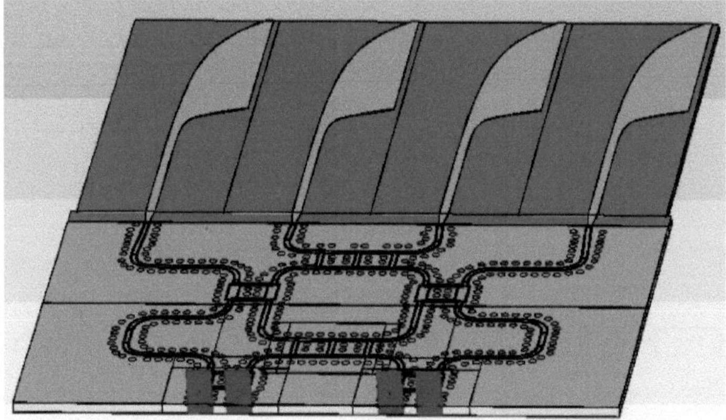

Fig. 3.32 Simulation setup for the Butler matrix with DETSA antenna.

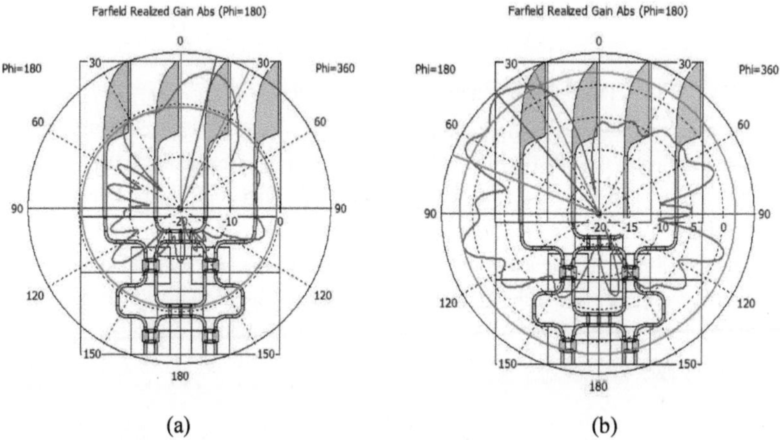

(a)   (b)

Fig. 3.33 Radiation pattern with excitation at (a) P1 and (b) P3.

Next, the design of the SP4T switch for controlling the port of excitation will be presented. The SP4T was designed using 0.15-μm GaN-on-SiC technology provided by FBH. A shunt-type configuration was adopted for the design of the SP4T switch. Fig. 3.34 shows the schematic of the SP4T switch.

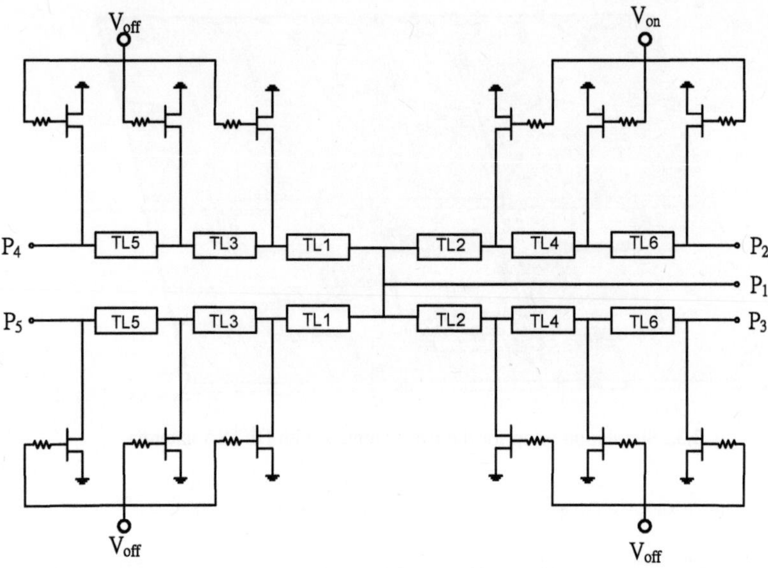

Fig. 3.34 Schematic of the SP4T switch.

The simulation procedure follows the design topology using traveling-wave concept [97]-[99]. The design specification of the SP4T switch will include an operating frequency in the $35 - 41$ GHz band, an insertion loss less than 4 dB, an isolation greater than 25 dB and an input P_{1dB} larger than 30 dBm.

For shunt-type switch designs, device with smaller gate peripheries are suitable in order to obtain insertion loss while devices with larger gate peripheries are beneficial for higher isolation. Therefore, the device size was chosen to be 2×50 μm. In order to improve the isolation, three shunt arms were designed to increase the level of isolation to meet the design specification of 25 dB.

The simulation results of the return losses, insertion loss and isolation are plotted in Fig. 3.35. Note that the small-signal was simulated when the signal flew from port 1 to port 2. Therefore, the supplied voltage was set to be 0.5 V for V_{on} and -3 V for V_{off}. From the simulated reflection coefficients (S_{11}, S_{22}, S_{33}, S_{44} and S_{55}), as port 1 and port 2 formed the path for signal transmission, both of them should exhibit a return loss better than 10 dB for proper connection with external circuitries. As for the isolated ports (port 3-5), the reflective type SP4T switch showed an isolated performance with a return loss of less than 10 dB as the signal transmission to these ports were limited. The characteristics including insertion loss (S_{21}) and isolation (S_{32}, S_{42}, S_{52}) are also plotted in Fig. 3.35 where the isolation between the ports match equally as the SP4T is designed in a symmetric topology.

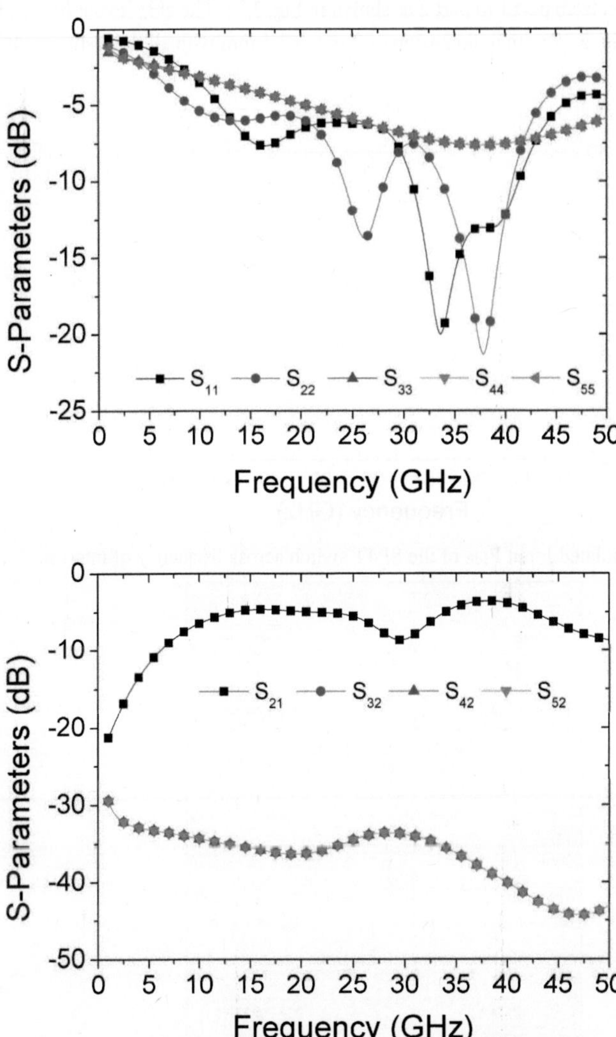

Fig. 3.35 Simulated small-signal results of the SP4T switch.

For the simulation results, the signal is transmitting from port 1 to port 2. Meanwhile, the other ports (port 3~port 5) are isolated. Since the presented results were extracted from simulation, the isolation performance was identical to each other. As observed, the small-signal characteristics of the SP4T switch meet the targeted specification well. The large-signal performance was also simulated using Keysight ADS simulator. Based on the statement given in chapter 2, the power was injected at the common input (P_1). Fig. 3.36 plots the input P_{1dB} of the SP4T switch as a function of frequency covering the frequency of interest. Note that the simulation results were performed with data transmission from port 1 to port 2 as shown in Fig. 3.34. The chip layout of the SP4T switch is shown in Fig. 3.37 with a dimension of 2 mm × 1.5 mm with all the testing pads and dicing streets included.

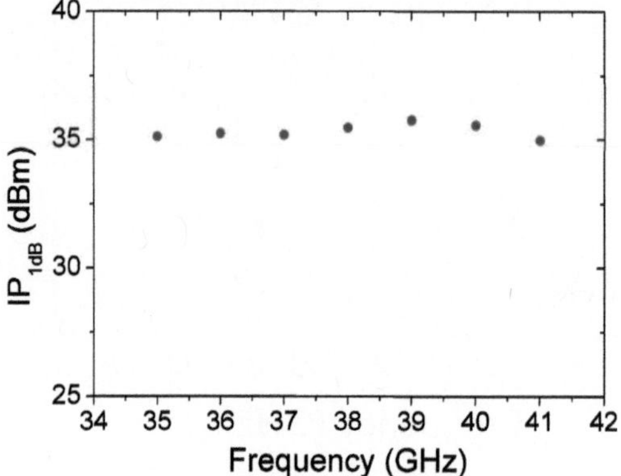

Fig. 3.36 Simulated Input P_{1dB} of the SP4T switch across frequency of interest.

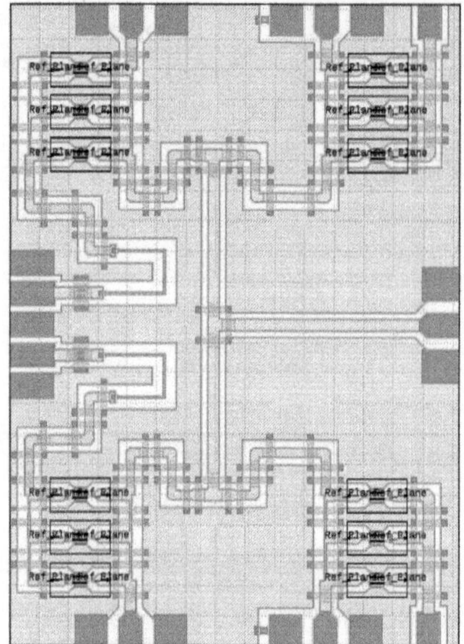

Fig. 3.37 Layout of the SP4T switch.

3.2.2. Technological implementation and measurement results

The fabricated Butler matrix using Rogers RO3010 substrate with CPW configuration is shown in Fig. 3.38. The small-signal performance was measured using Keysight N5227B PNA 4-ports measurements. For instance, P3, P4, P7, P8 were terminated with a 50 Ω while characterizing the other 4 ports. The measurement results are summarized in Table 3.5. Comparing with the simulated results shown in Table 3.4, the insertion losses of each ports have increased by around 1 dB as some of the phase differences between the ports are different from the simulation results. This may be referred to the slight difference of the extended signal line for connection with the endlaunch connectors.

Fig. 3.37 Layout of the SP4T switch.

Table 3.5. Measured performance of the Butler matrix.

Input / Output	Port 1 (1L)		Port 2 (2R)		Port 3 (2L)		Port 4 (1R)	
	Phase (°)	IL (dB)	Phase (°)	IL (dB)	Phase (°)	IL (dB)	Phase (°)	IL (dB)
Port 5	48	9.3	134	9.1	92	7.9	184	5.9
Port 6	86	9	3	9.6	222	5.8	135	9.3
Port 7	137	8.7	226	5.4	10	10.2	91	8.6
Port 8	177	6.1	83	7.6	137	9.0	44	9.1

The measurement shows a promising results for further integration with the SP4T switch.

3.3. Module integration and characterization of complete module

In this section, two transceiver modules will be presented using the designed SPDT switches, LNPA and DETSA antenna. The transceiver modules will be mounted on a Rogers RO3010 substrate and connected to the leads on the substrate by wire bonding. Prior to the hybrid integration of the complete transceiver module, an analysis with the bonding wires was done for estimation of the parasitic effects. Fig. 3.38 shows the test structure of the wire bonds. In this analysis, different lengths of wire bonds were applied to connect the signal lines on the two sides of the carrier. This experiment was expected beneficial for estimating the parasitic effects induced by the wire bonds in factors of length. Although the realistic implementation of wire bond connection has a height difference between the mounted chips and carrier, this factor was neglected during the analysis for reducing the complexity of extraction and external issues.

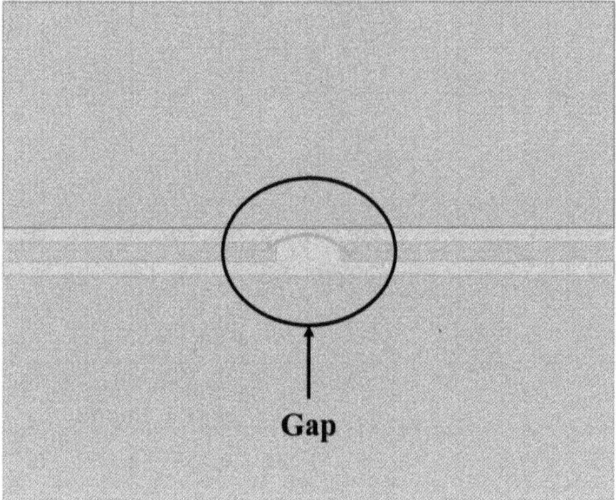

Figure 3.38. Test structure for wire bonds.

The characterization of wire bonds was reported in [100], the equivalent circuit model is shown in Fig. 3.39.

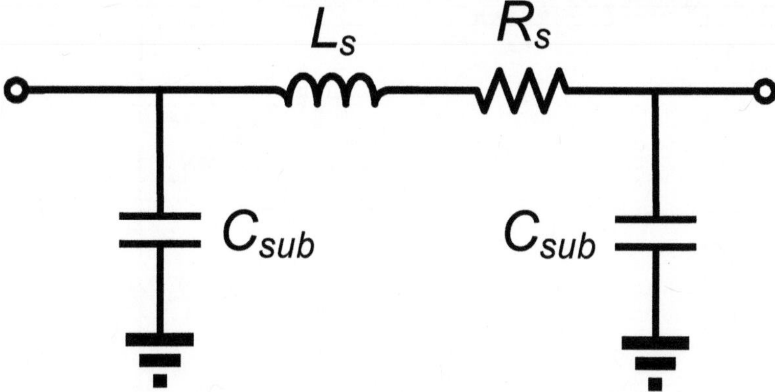

Figure 3.39. Equivalent circuit for wire bond interconnect.

Measurement results of wire bond and ribbon bond connecting with different gap length are plotted in Fig. 3.40 and 3.41.

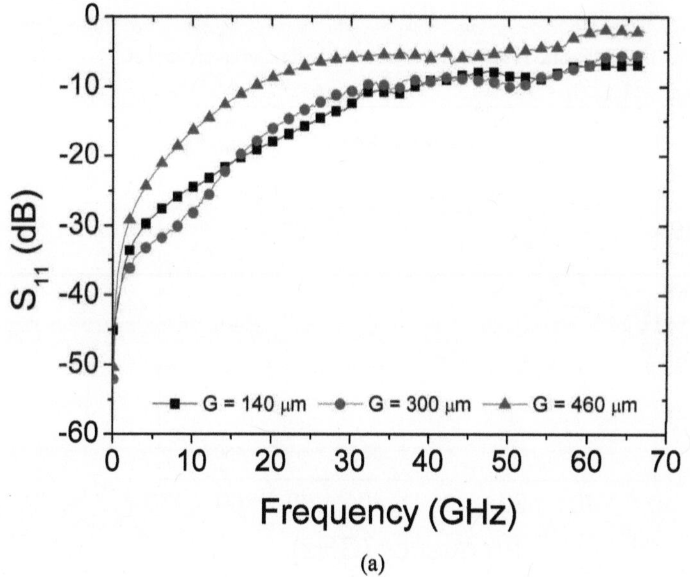

Frequency (GHz)

(a)

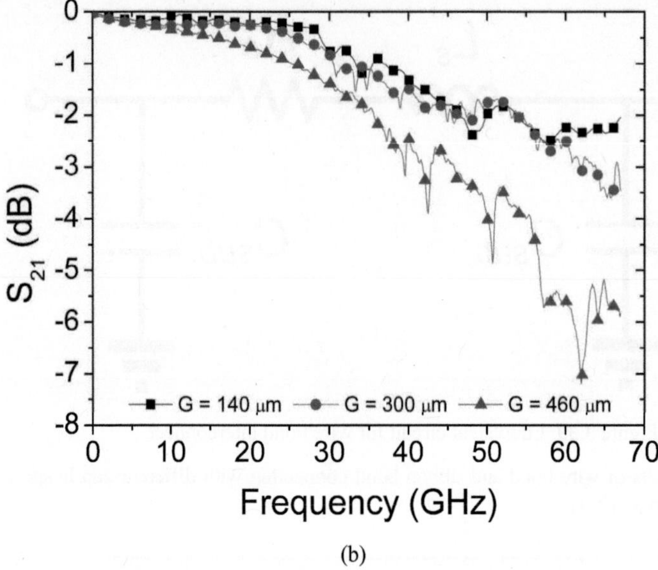

(b)

Figure 3.40. Measured (a) S_{11} and (b) S_{21} for ball bond interconnect.

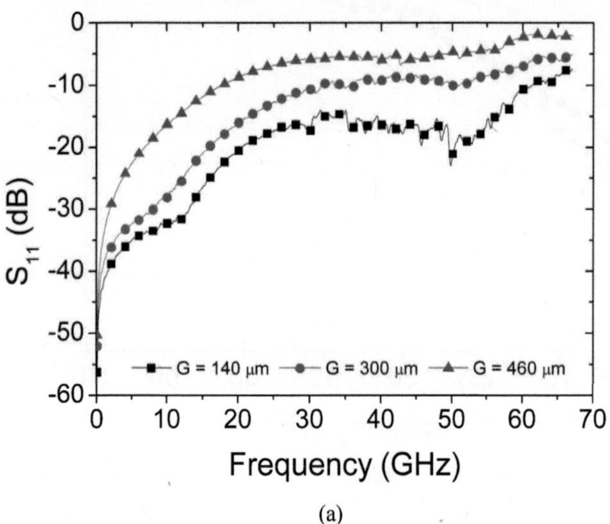

(a)

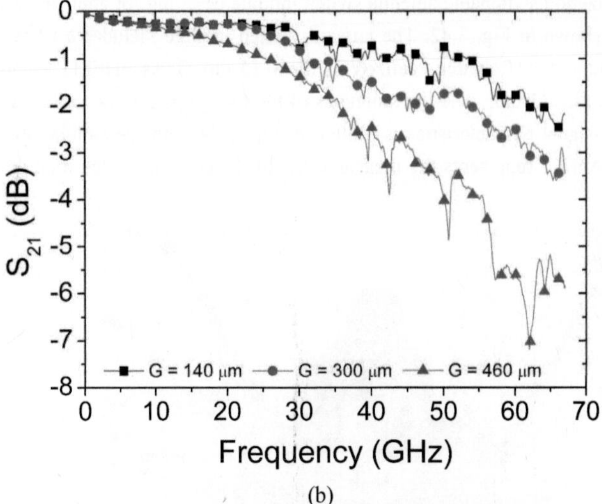

(b)

Figure 3.41. Measured (a) S$_{11}$ and (b) S$_{21}$ for ribbon bond interconnect.

With the measurement results, the test structure has been de-embedded to the bonding pads for extraction of the parasitic effects induced by the wire bonds. The parasitic effects of the ball bonds can be expressed as:

$$L_s = 0.87\frac{\text{pH}}{\mu\text{m}}\cdot \text{Gap}(\mu\text{m})-10.69\text{pH} \tag{3.3}$$

$$R_s = 0.0014\frac{\Omega}{\mu\text{m}}\cdot \text{Gap}(\mu\text{m})+0.216\ \Omega \tag{3.4}$$

$$C_{sub} = 0.0049\frac{\text{fF}}{\mu\text{m}}\cdot \text{Gap}(\mu\text{m})+12.539\ \text{fF} \tag{3.5}$$

As for the ribbon bond interconnects, the extracted expression for the parasitics are shown as follows:

$$L_s = 0.51\frac{\text{pH}}{\mu\text{m}}\cdot \text{Gap}(\mu\text{m})-7.23\ \text{pH} \tag{3.6}$$

$$R_s = 0.0011\frac{\Omega}{\mu\text{m}}\cdot \text{Gap}(\mu\text{m})+0.105\ \Omega \tag{3.7}$$

$$C_{sub} = 0.0048\frac{\text{fF}}{\mu\text{m}}\cdot \text{Gap}(\mu\text{m})+13.435\ \text{fF} \tag{3.8}$$

As the ribbon bond has a larger width comparing with the ball bond, a lower equivalent inductance is extracted based on the measurement results. The estimation of parasitic effects helps to work with the pre-matching of the module integration.

The fabricated polarization switchable antenna switch module targeting for application at Ka-band frequencies is shown in Fig. 3.42. The antenna switch module includes a LNPA from GNQ-06 process and a SPDT switch realized in the 0.15-μm GaAs pHEMT technology provided by WIN Semiconductor. The interconnects of the two chips are shown in Fig. 3.43. The measured small-signal characteristics is plotted in Fig. 3.44 with the simulation results included. Note that ANT 1 represents the measured S_{11} of the antenna switch module while ANT 1 is excited.

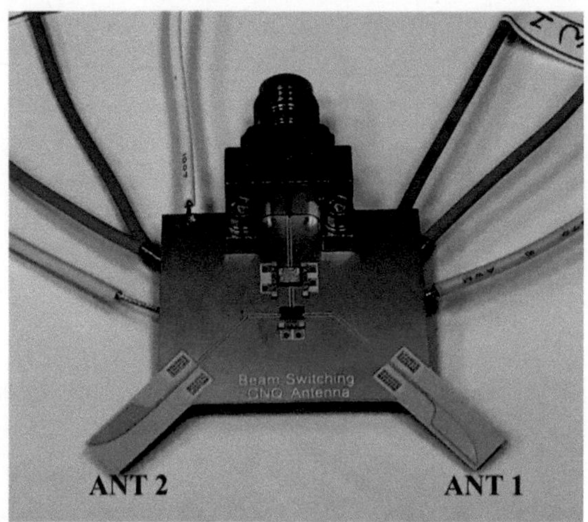

Figure 3.42. Fabricated polarization switchable antenna switch module.

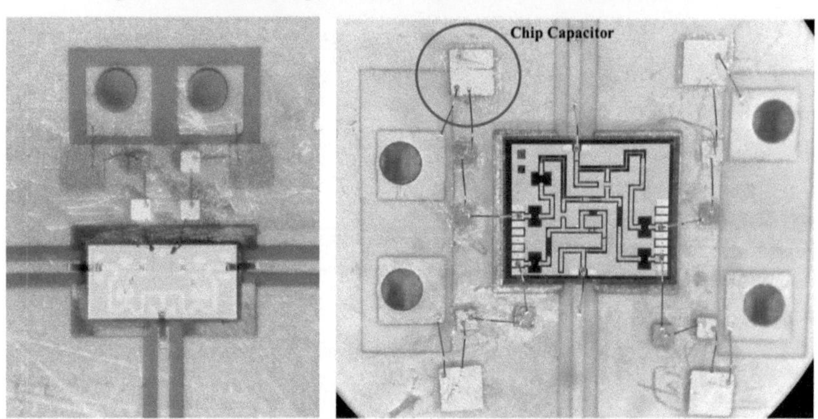

Figure 3.43. Interconnects of the mounted chips.

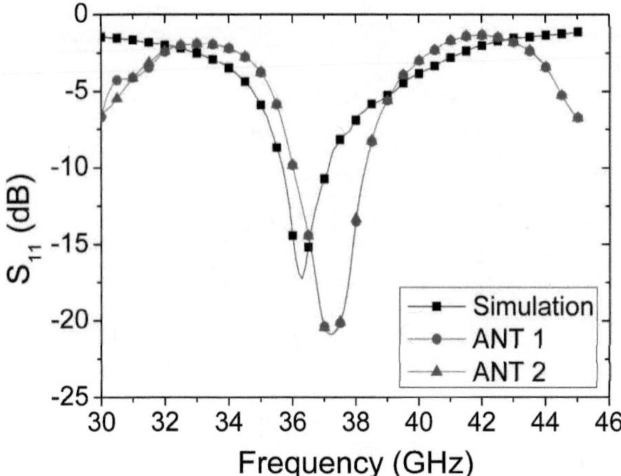

Figure 3.44. Small-signal characteristics of the polarization switchable antenna switch module.

An over the air test was performed to characterize the radiation pattern of the antenna switch module with free space transmission. Fig. 3.45 shows the measurement setup for E-plane radiation characterization. During the measurement, the antenna switch module was set in the receiving mode to receive the signal transmitted from the horn antenna. The Keysight E8257D signal generator was applied for the generation of CW signal. The antenna switch module was connected to the Anritsu MS2840A spectrum analyzer for monitoring the received signal. The power supply was applied to control the signal path by setting the switch at different operation mode. A free space path loss is calculated to be 73.58 dB at 38 GHz. With an input power of 10 dBm set for the signal generator, the normalized measured radiation pattern is plotted in Fig. 3.46.

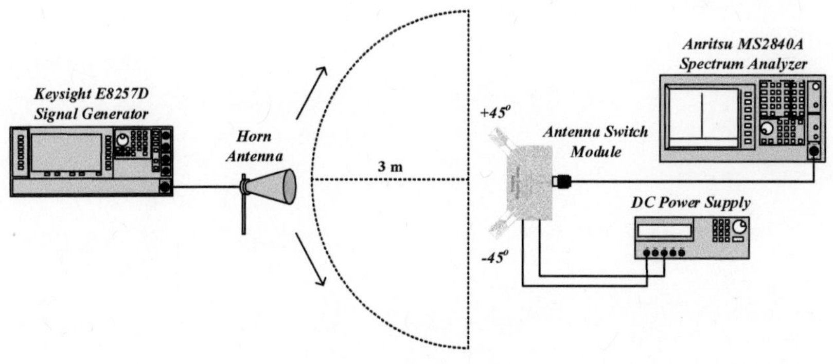

Figure 3.45. Measurement setup for over the air test.

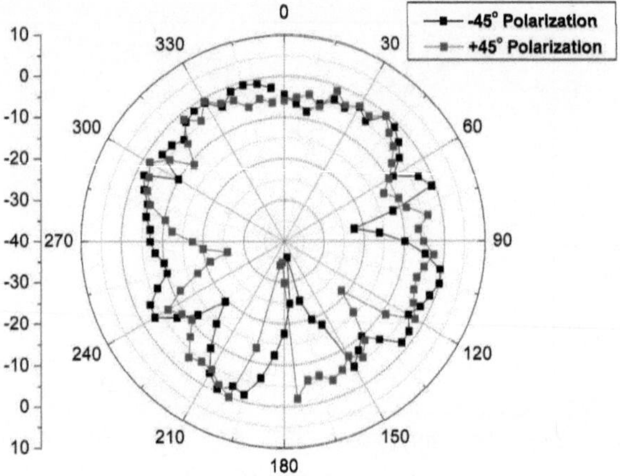

Figure 3.46. The measured radiation pattern of the antenna switch module under two states of operation.

The measurement results shown in Fig. 3.46 with different angle of polarization showed a symmetric pattern as the antenna elements were placed symmetrically. The measurement results meet the pattern measured with the single DETSA antenna. The beam steering characteristic is not that obvious because for signal element, the radiation beamwidth is relatively large.

A transceiver module using the design concept of LNPA with two SPDT switches as shown in Fig. 2.2 is fabricated using Rogers RO3010 substrate. The fabricated transceiver module is shown in Fig. 3.47. Similar interconnect approaches were applied to the transceiver module as compared to the previously discussed polarization switchable antenna module. The measured small-signal characteristics are plotted as Fig. 3.48, where transmitting and receiving mode are both characterized.

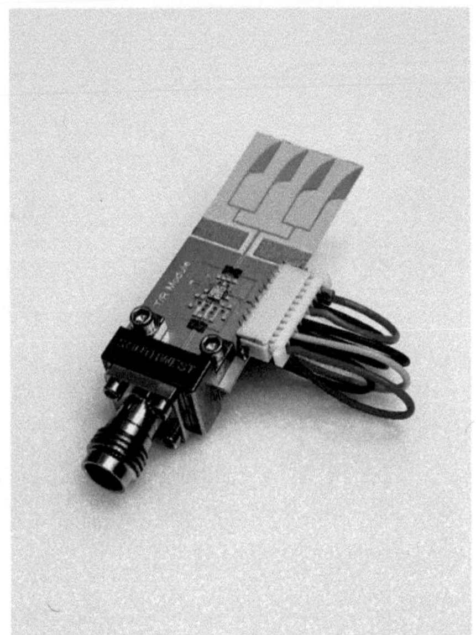

Figure 3.47. Fabricated transceiver using LNPA approach.

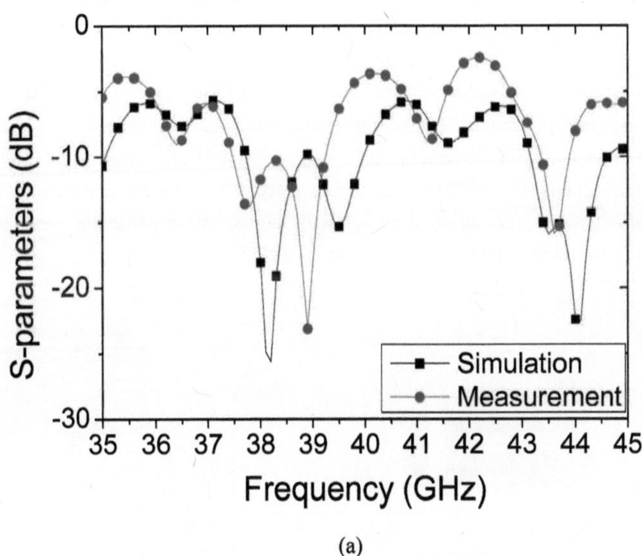

(a)

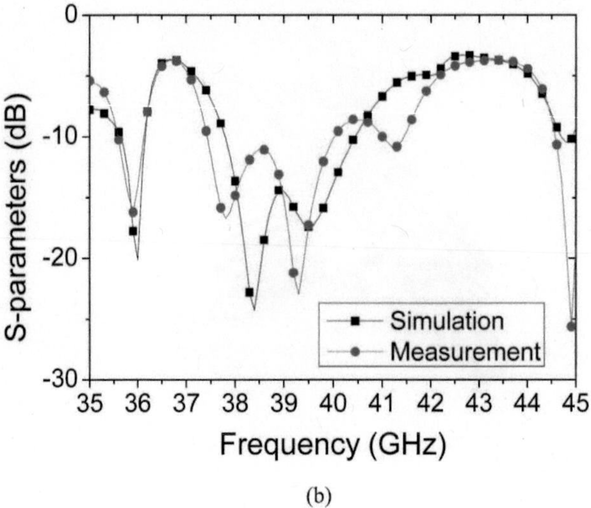

(b)

Figure 3.48. Measured input return loss of the transceiver module in (a) receiving and (b) transmitting mode.

The measured input return loss showed a nice agreement with the simulation results using CST Studio Suite. This proves that the transceiver module can properly operate at the desired frequencies for over the air test. An over the air test was then performed to characterize the radiation pattern of the transceiver module. The radiation pattern of the transceiver module was characterized under transmitting mode. The measurement setup is shown in Fig. 3.49, where the distance was set to be 1 m with a free space path loss of 64.04 dB. In this setup, the horn antenna was connected to the Anritsu MS2840A spectrum analyzer for monitoring the receiving signal. Fig. 3.50 plots the measured radiation pattern with normalization while the fabricated transceiver module is delivering a transmitted power of 20 dBm.

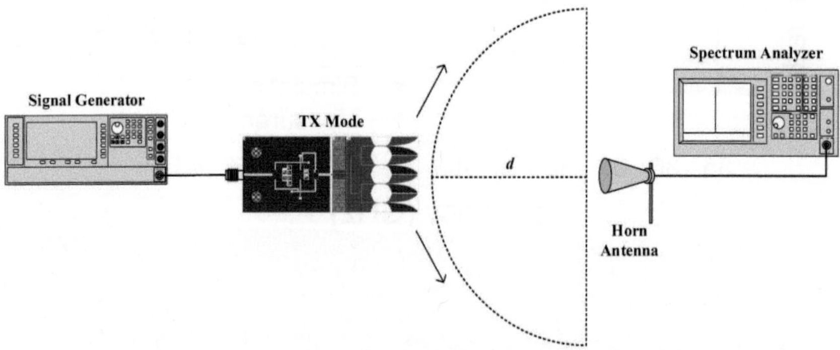

Figure 3.49. Point-to-Point Wireless Transmission Test Bench.

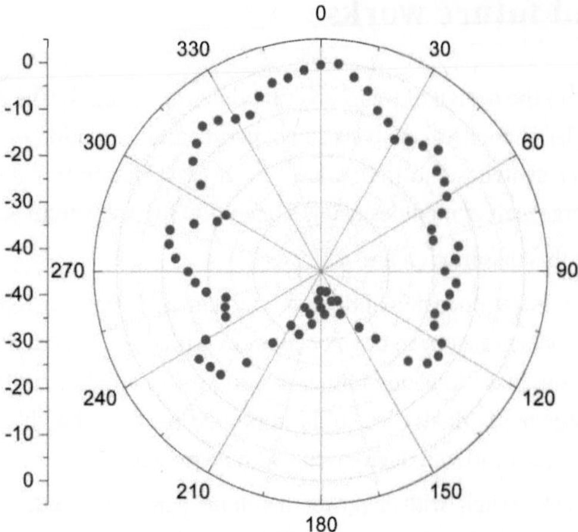

Figure 3.50. Measured radiation pattern under transmitting mode.

The measured radiation pattern shows the capability of the transceiver module for operating at transmitting mode. With the evaluation of design specifications of the implemented RF front-end circuits including the antenna array, it evidences that the design concept using a LNPA for the transceiver module is promising for future communication systems.

4. Conclusion and future works

This dissertation is devoted to the design of transceiver modules targeting for the application at 5G NR FR2 frequency bands. Theoretical analysis and experimental validation of the RF front-end circuits including SPDT switches and LNPA have been presented in this dissertation. In addition, the design of the transceiver modules using Rogers RO3010 substrate is presented.

The major achievements in this dissertation are:

(a) A design approach for isolation bandwidth improvement of SPDT switches using the impedance transform concept has been developed and verified for Ka-band applications. A theoretical analysis on possible bandwidth expansion using a radial stub concept was performed. Such arrangements finally led to an improved isolation bandwidth of SPDT switch compared to the conventional open-circuit stub approach.

(b) A X- to Ka-band SPDT switch with a power handling capability better than 3 W is demonstrated using a 0.15-μm GaAs pHEMT technology. This is a world-wide competitive property. A wide operating bandwidth was achieved using the series-shunt configuration. Theoretical analysis has been performed to analyze the intrinsic effect of the devices. A stacked-FET configuration was adopted for improving power handling capability by analyzing the pros and cons for the number of transistors in stack and a proper selection of the device size. The power handling capability of the series-shunt SPDT switch utilizing stacked-FET configuration shows a comparable result with previously reported high power switches in other technologies.

(c) For operation at millimeter-wave frequencies, the roll-off of the MAG has been an issue for realizing wideband or multi-band amplifiers with similar performance. A dual-band power amplifier targeting for the carrier aggregation application at 5G NR FR2 n257 and n260 bands is presented. With the utilization of stacked-FET configuration for the mitigation of the MAG roll-off, the dual-band amplifier achieves an identical performance at both 28 and 38 GHz. The experimental has demonstrated only a 0.5-dB difference of the small-signal gain measured at 28 and 38 GHz. The experimental results have shown great potential for integration with the dual-band transceiver.

(d) Successful implementation of a two-stage LNPA using FBH 0.15-μm GaN-on-SiC technology with only 3 dB back-off from P_{sat} to achieve an EVM less than 5% was presented for 5G NR FR2 application. An analysis for the noise characteristics of the GaN HEMT applied for the low-noise driver stage design was performed. It turned out that a trade-off between the MAG and NF_{min} was necessary. This has been optimized to achieve reasonable noise and gain performance. The power stage was matched to the optimum impedance for maximum output power. Quite some effort has been devoted to a proper design of the interstage matching network was designed using a low-Q matching topology composed of a piece of series transmission line and a DC blocking capacitor. Such an arrangement helped to significantly reduce the matching loss for achieving a higher transducer gain. Additionally, as compared to conventional LC networks, this also reduces the layout dimension. The linearity was characterized using 64-QAM modulation signals and compared to the previously published CMOS and GaAs power amplifiers. Only 2.2-dB back-off from the P_{sat} of the LNPA was needed to maintain operation at the EVM level of 5%. The experimental results evidence that GaN technology is promising for the next generation communication system for linear operation with high output power.

(e) A polarization switchable antenna switch module was integrated with the design of RF front-end circuits. An analysis on the parasitic effects of the wire bonds has made the transceiver integration much easier. The polarization switchable antenna switch module is believed possible as scalable approach for larger antenna array integration with a lower cost.

(f) A transceiver module with the implementation of LNPA with two reflective SPDT switches was designed and fabricated. Such an arrangement can effectively reduce the parasitic effects induced by the wire bonds, which is due to the reduced number of amplifiers needed for integration. The compact transceiver module is considered to be able for scaling up to a larger phased-array system for high power applications such as base station communication technologies. The lower complexity also reduces costs and integration efforts and thus enables promising concepts for future 5G NR FR2 communication systems.

(g) As communication systems are tending to use higher order of modulation schemes for the next generation applications, the stringent requirements make transceiver integration tougher comparing with the old days. For mature processing technologies such as CMOS and GaAs, the back-off from P_{sat} were normally set at 10 and 6 dB, where the amplifiers were operating at low PAE region. The GaN-based design has shown a good linearity performance at less than 3-dB back-off from P_{sat}. Such low back-off value was mainly attributed to the higher Schottky barrier of the gate module developed in FBH 0.15-μm GaN-on-SiC technology.

Targeting for the application of the next generation communication systems, there are still many design issues that have to be addressed:

(a) While targeting for high power applications, the large amount of heat induced during operation is a complicated issue to solve. A potential solution for spreading out the dissipated heat can be solved using thermally highly conductive and electrically isolating substrates such as aluminum nitride (AlN). To directly connect the chips with AlN substrate, hot-via technologies are expected to achieve excellent heat dissipation with low loss interconnects between chip and substrate for applications up to 100 GHz. Further investigations on the new packaging techniques are expected to be done in the future.

(b) As highly reliable RF front-end circuits are key components for applications such as space communication and electronic countermeasure. Systematic design approach improving thermal reliability of the RF front-end circuits shall also be adopted. For instance, stacked-FET configuration has been reported with a nice thermal reliability using GaN technology previously. Design approaches using layout modification is believed to change thermal effect induced during operation. Therefore, analysis on the design of active components with thermal reliability must be investigated.

5. References

[1] M. H. Alsharif, A. H. Kelechi, M. A. Albreem, S. A. Chaudhry, M. S. Zia and S. Kim, "Sixth Generation (6G) Wireless Networks: Vision, Research Activities, Challenges and Potential Solutions," *MDPI Sensors*, 2020, 12, 676.

[2] A. Ghosh, A. Maeder, M. Baker and D. Chandramouli, "5G Evolution: A View on 5G Cellular Technology Beyond 3GPP Release 15," *IEEE Access*, vol. 7, pp. 127639-127651, 2019.

[3] IEEE Future Networks, "IEEE 5G and Beyond Technology Roadmap White Paper." Available online: https://www.futurenetworks.ieee.org/roadmap (accessed on 15 November 2019)

[4] E. J. Candes, J. Romberg and T. Tao, "Robust uncertainty principles: exact signal reconstruction from highly incomplete frequency information," *IEEE Trans. Info. Theory*, vol. 52, no. 2, pp. 489-509, Feb. 2006.

[5] Qualcomm, "Global update on spectrum for 4G and 5G," Dec. 8, 2020. [Online] Available: https://www.qualcomm.com/media/documents/files/spectrum-for-4g-and-5g.pdf

[6] European Telecommunications Standards Institute, "5G NR Base Station (BS) radio transmission and reception," 3GPP TS 38.104 Version 15.2.0 Release 15, Jul., 2018. [Online] Available: https://www.etsi.org/deliver/etsi_ts/138100_138199/138104/15.02.00_60/ts_138104v150200p.pdf

[7] Y. Golovachev, A. Etinger, G. A. Pinhasi and Y. Pinhasi, "Millimeter Wave High Resolution Radar Accuracy in Fog Conditions—Theory and Experimental Verification," *MDPI Sensors*, 18, 2148, 2018.

[8] J. Krupka, "Microwave Measurements of Electromagnetic Properties of Materials," *MDPI Materials*, 14, 5097, 2021.

[9] Y. Yashchyshyn, P. Bajurko, J. Sobolewski, P. Sai, A. Przewłoka, A. Krajewska, P. Prystawko, M. Dub, W. Knap, S. Rumyantsev and G. Cywiński, "Graphene/AlGaN/GaN RF Switch," *MDPI Micromachines*, 12, 1343, 2021.

[10]   F. Sheikh, M. Alissa, A. Zahid, Q. H. Abbasi and T. Kaiser, "Atmospheric Attenuation Analysis in Indoor THz Communication Channels," in *Proc. 2019 IEEE Int. Symp. Ant. Prop. USNC-URSI Radio Sci. Meeting*, Atlanta, GA, USA, 2019, pp. 2137-2138.

[11]   H. Khatibi, S. Khiyabani and E. Afshari, "A 173 GHz Amplifier With a 18.5 dB Power Gain in a 130 nm SiGe Process: A Systematic Design of High-Gain Amplifiers Above $f_{max}/2$," *IEEE Trans. Microw. Theory Techn.*, vol. 66, no. 1, pp. 201-214, Jan. 2018.

[12]   M. Božanić and S. Sinha, " Emerging Transistor Technologies Capable of Terahertz Amplification: A Way to Re-Engineer Terahertz Radar Sensors," *MDPI Sensor*, 19, 2454, 2019.

[13]   H. Wang *et al.* (Aug. 2021). Power Amplifiers Performance Survey 2000-Present. [Online]. Available: https://gems.ece.gatech.edu/PA_survey.html

[14]   O. Momeni, "A 260 GHz amplifier with 9.2 dB gain and -3.9 dBm saturated power in 65 nm CMOS," in *Proc. IEEE Solid-State Circuits Conf. Dig.*, Feb. 2013, pp. 140–141.

[15]   D. Simic and P. Reynaert, "Analysis and Design of Lossy Capacitive Over-Neutralization Technique for Amplifiers Operating Near f_{MAX}," *IEEE Trans. Circuits Syst. I: Regul. Pap*, vol. 68, no. 5, pp. 1945-1955, May 2021.

[16]   D. P. Nguyen, A. N. Stameroff and A. Pham, "A 1.5–88 GHz 19.5 dBm output power triple stacked HBT InP distributed amplifier," in *Proc. IEEE MTT-S Int. Microw. Symp. Dig.*, Jun. 2017, pp. 20-23.

[17]   J. Essing, R. Mahmoudi, Y. Pei and A. van Roermund, "A fully integrated 60GHz distributed transformer power amplifier in bulky CMOS 45nm," in *Proc. 2011 IEEE Radio Freq. Integr. Circuits Symp.*, July 2011, pp. 1-4.

[18]   "Millimeter-Wave (MMW) Radio Transmission: Atmospheric Propagation, Link Budget and System Availability, " *Light Pointe White Paper Series*, 2010.

[19]   Rappaport, T. S., R. Mayzus, Y. Azar, K. Wang, G. N. Wong, J. K. Schulz, M. Samimi, and F. Gutierrez, "Millimeter wave mobile communications for 5G cellular: It will work!," *IEEE Access*, vol. 1, 335–349, 2013.

[20]   Z. Jiang, P. Zheng, Y. Qiu, H. Zhu, C. Ding and G. Wei, "Mechanically Beam-steering of Horn Antenna Using Non-uniform Flexible Meta-Surface," in *Proc. 2020 IEEE 6th Int. Conf. Computer Comm. (ICCC)*, 2020, pp. 463-466.

[21]   A. Sutinjo and M. Okoniewski, "A Surface Wave Holographic Antenna for Broadside Radiation Excited by a Traveling Wave Patch Array," *IEEE Trans. Antennas Propag.*, vol. 59, no. 1, pp. 297-300, Jan. 2011.

[22]   A. Artemenko, A. Mozharovskiy, A. Maltsev, R. Maslennikov, A. Sevastyanov and V. Ssorin, "Experimental Characterization of E-Band Two-Dimensional Electronically Beam-Steerable Integrated Lens Antennas," *IEEE Antennas Wirel. Propag. Lett.*, vol. 12, pp. 1188-1191, 2013.

[23]   B. Schaer, K. Rambabu, J. Bornemann and R. Vahldieck, "Design of reactive parasitic elements in electronic beam steering arrays," *IEEE Trans. Antennas Propag.*, vol. 53, no. 6, pp. 1998-2003, June 2005.

[24]   G. Wu, S. Qu and S. Yang, "Wide-Angle Beam-Scanning Reflectarray With Mechanical Steering," *IEEE Trans. Antennas Propag.*, vol. 66, no. 1, pp. 172-181, Jan. 2018.

[25]   R. Lu, C. Weston, D. Weyer, F. Buhler, D. Lambalot and M. P. Flynn, "A 16-Element Fully Integrated 28-GHz Digital RX Beamforming Receiver," *IEEE J. Solid-State Circuits*, vol. 56, no. 5, pp. 1374-1386, May 2021.

[26]   P. Kyösti, "Correlation of Hybrid Beamforming Arrays in the Context of OTA Testing," *IEEE Antennas Wirel. Propag. Lett.*, vol. 19, no. 4, pp. 671-675, April 2020.

[27]   E. Carrasco and J. Perruisseau-Carrier, "Reflectarray Antenna at Terahertz Using Graphene," *IEEE Antennas Wirel. Propag. Lett.*, vol. 12, pp. 253-256, 2013.

[28]   K. Trzebiatowski, M. Rzymowski, L. Kulas and K. Nyka, "Simple 60 GHz Switched Beam Antenna for 5G Millimeter-Wave Applications," *IEEE Antennas Wirel. Propag. Lett.*, vol. 20, no. 1, pp. 38-42, Jan. 2021.

[29]   V. F. Fusco and N. Buchanan, "Retrodirective Array Performance in the Presence of Near Field Obstructions," *IEEE Trans. Antennas Propag.*, vol. 58, no. 3, pp. 982-986, March 2010.

[30]   J. Reis, R. Z. Al-Daher, N. Copner, R. Caldeirinha and T. Fernandes, "Two-dimensional antenna beamsteering using metamaterial transmitarray," in *Proc. 2015 9th Euro. Conf. Antennas Propag. (EuCAP)*, 2015, pp. 1-5.

[31]   W. Roh *et al.*, "Millimeter-wave beamforming as an enabling technology for 5G cellular communications: Theoretical feasibility and prototype results," *IEEE Commun. Mag.*, vol. 52, no. 2, pp. 106–113, Feb. 2014.

[32]   B. Sadhu *et al.*, "A 28-GHz 32-Element TRX Phased-Array IC With Concurrent Dual-Polarized Operation and Orthogonal Phase and Gain Control for 5G Communications," *IEEE J. Solid-State Circuits*, vol. 52, no. 12, pp. 3373-3391, Dec. 2017.

[33]   E. Cohen, M. Ruberto, M. Cohen, O. Degani, S. Ravid, and D. Ritter, "A CMOS bidirectional 32-element phased-array transceiver at 60 GHz with LTCC antenna," *IEEE Trans. Microw. Theory Techn.*, vol. 61, no. 3, pp. 1359–1375, Mar. 2013.

[34]   A. Valdes-Garcia *et al.*, "A fully-integrated dual-polarization 16-element W-band phased-array transceiver in SiGe BiCMOS," in *Proc. IEEE Radio Freq. Integr. Circuits Conf. (RFIC)*, Jun. 2013, pp. 375–378.

[35]   H.-T. Kim *et al.*, "A 28 GHz CMOS direct conversion transceiver with packaged antenna arrays for 5G cellular system," in *Proc. IEEE Radio Freq. Integr. Circuits Conf. (RFIC)*, Honolulu, HI, USA, Jun. 2017, pp. 69–72.

[36]   S. Shahramian, M. J. Holyoak, and Y. Baeyens, "A 16-element W-band phased array transceiver chipset with flip-chip PCB integrated antennas for multi-gigabit data links," in *Proc. IEEE Radio Freq. Integr. Circuits Conf. (RFIC)*, May 2015, pp. 27–30.

[37]   A. Townley *et al.*, "A 94 GHz 4TX-4RX phased-array for FMCW radar with integrated LO and flip-chip antenna package," in *Proc. IEEE Radio Freq. Integr. Circuits Conf. (RFIC)*, May 2016, pp. 294–297.

[38]   K. Kibaroglu, M. Sayginer, and G. M. Rebeiz, "An ultra low-cost 32-element 28 GHz phased-array transceiver with 41 dBm EIRP and 1.0–1.6 Gbps 16-QAM link at 300 meters," in *Proc. IEEE Radio Freq. Integr. Circuits Conf. (RFIC)*, Honolulu, HI, USA, Jun. 2017, pp. 73–76.

[39]   K. Kibaroglu, M. Sayginer, and G. M. Rebeiz, "A quad-core 28–32 GHz transmit/receive 5G phased-array IC with flip-chip packaging in SiGe BiCMOS," in *Proc. IEEE Int. Microw. Symp. (IMS)*, Jun. 2017, pp. 1892–1894.

[40]   C. -N. Chen *et al.*, "38-GHz Phased Array Transmitter and Receiver Based on Scalable Phased Array Modules With Endfire Antenna Arrays for 5G MMW Data Links," *IEEE Trans. Microw. Theory Techn.*, vol. 69, no. 1, pp. 980-999, Jan. 2021.

[41]   K. Takahashi, S. Okamura, X. Wang, M. Tahara and I. Sakagami, "A capacitance compensated high isolation and low insertion loss series PIN diode SPDT switch," in *Proc. 2011 41st Euro. Microw. Conf.*, 2011, pp. 583-586.

[42]   J. Park, W. Lee and S. Hong, "A Small-Size K-Band SPDT Switch Using Alternate CMOS Structure With Resonating Inductor Matching," *IEEE Microw. Wirel. Compon. Lett.*, vol. 30, no. 11, pp. 1093-1096, Nov. 2020.

[43]   K.-Y. Lin, Wen-Hua Tu, Ping-Yu Chen, Hong-Yeh Chang, Huei Wang and Ruey-Beei Wu, "Millimeter-wave MMIC passive HEMT switches using traveling-wave concept," *IEEE Trans. Microw. Theory Techn.*, vol. 52, no. 8, pp. 1798-1808, Aug. 2004.

[44]   L. Zhao, W. Liang, J. Zhou and X. Jiang, "Compact 35–70 GHz SPDT Switch With High Isolation for High Power Application," *IEEE Microw. Wirel. Compon. Lett.*, vol. 27, no. 5, pp. 485-487, May 20174.

[45]   H. Mizutani, N. Funabashi, M. Kuzuhara and Y. Takayama, "Compact DC-60-GHz HJFET MMIC switches using ohmic electrode-sharing technology," *IEEE Trans. Microw. Theory Techn.*, vol. 46, no. 11, pp. 1597-1603, Nov. 1998.

[46]   M. J. Schindler and A. Morris, "DC-40 GHz and 20-40 GHz MMIC SPDT Switches," *IEEE Trans. Microw. Theory Techn.*, vol. 35, no. 12, pp. 1486 1493, Dec 1987.

[47]   B. Min and G. M. Rebeiz, "Ka-Band Low-Loss and High-Isolation Switch Design in 0.13-μm CMOS," *IEEE Trans. Microw. Theory Techn.*, vol. 56, no. 6, pp. 1364-1371, June 2008.

[48]   J. Lee *et al.*, "Low Insertion-Loss Single-Pole–Double-Throw Reduced-Size Quarter-Wavelength HEMT Bandpass Filter Integrated Switches," *IEEE Trans. Microw. Theory Techn.*, vol. 56, no. 12, pp. 3028-3038, Dec. 2008.

[49]   Yoshihiro Tsukahara, Hirotaka Amasuga, Seiki Goto, Tomoki Oku and Takahide Ishikawa, "60GHz High Isolation SPDT MMIC switches using shunt pHEMT resonator," in *Proc. 2008 IEEE MTT-S Int. Microw. Symp. Dig.*, 2008, pp. 1541-1544.

[50]   A. Ç. Ulusoy et al., "A Low-Loss and High Isolation D-Band SPDT Switch Utilizing Deep-Saturated SiGe HBTs," *IEEE Microw. Wirel. Compon. Lett.*, vol. 24, no. 6, pp. 400-402, June 2014.

[51]   D. P. Nguyen, A. Pham and F. Aryanfar, "A K-Band High Power and High Isolation Stacked-FET Single Pole Double Throw MMIC Switch Using Resonating Capacitor," *IEEE Microw. Wirel. Compon. Lett.*, vol. 26, no. 9, pp. 696-698, Sept. 2016.

[52]   S. Kaleem, J. Kühn, R. Quay and M. Hein, "A high-power Ka-band single-pole single-throw switch MMIC using 0.25 μm GaN on SiC," in *Proc. 2015 IEEE Radio Wirel. Symp. (RWS)*, 2015, pp. 132-134.

[53]   K. Lee, S. Choi and C. Kim, "A 25–30-GHz Asymmetric SPDT Switch for 5G Applications in 65-nm Triple-Well CMOS," *IEEE Microw. Wirel. Compon. Lett.*, vol. 29, no. 6, pp. 391-393, June 2019.

[54]   P. Park, D. H. Shin, and C. P. Yue, "High-Linearity CMOS T/R switch design above 20 GHz using asymmetrical topology and AC-floating bias," *IEEE Trans. Microw. Theory Techn.*, vol. 57, no. 4, pp. 948–956, Apr. 2009.

[55]   Kun-You Lin, Yu-Jiu Wang, Dow-Chih Niu and Huei Wang, "Millimeter-wave MMIC single-pole-double-throw passive HEMT switches using impedance-transformation networks," *IEEE Trans. Microw. Theory Techn.*, vol. 51, no. 4, pp. 1076-1085, April 2003.

[56]   P. Mei *et al.*, "Single-Pole Double-Throw Switch Using Stacked-FET Configuration at Millimeter Wave Frequencies," in *Proc. 2018 Asia-Pacific Microw. Conf. (APMC)*, 2018, pp. 791-793.

[57]   H. Dabag *et al.*, "Analysis and Design of Stacked-FET Millimeter-Wave Power Amplifiers," *IEEE Trans. Microw. Theory Techn.*, vol. 61, no. 4, pp. 1543-1556, April 2013.

[58]   Y. Kim and Y. Kwon, "Analysis and Design of Millimeter-Wave Power Amplifier Using Stacked-FET Structure," *IEEE Trans. Microw. Theory Techn.*, vol. 63, no. 2, pp. 691-702, Feb. 2015.

[59]   G. van der Bent, P. de Hek and F. E. van Vliet, "Design Procedure for Integrated Microwave GaAs Stacked-FET High-Power Amplifiers," *IEEE Trans. Microw. Theory Techn.*, vol. 67, no. 9, pp. 3716-3731, Sept. 2019.

[60]   C.-G. Yuan *et al.*, "0.15 Micron Gate 6-inch pHEMT Technology by Using I-Line Stepper," in *Proc. CS MANTECH Conf.*, Tampa, Florida, USA, May 2009.

[61]   C. Chiang and H. Hsu, "Design and implementation of multi-octave low-noise power amplifier (LNPA) using HIFET configuration," in *Proc. 2012 Asia Pacific Microw. Conf. (APMC)*, 2012, pp. 956-958.

[62]   S. Masuda, T. Ohki and T. Hirose, "Very Compact High-Gain Broadband Low-Noise Amplifier in InP HEMT Technology," *IEEE Trans. Microw. Theory Techn.*, vol. 54, no. 12, pp. 4565-4571, Dec. 2006.

[63]   Y. Yu, W. Hsu and Y. E. Chen, "A Ka-Band Low Noise Amplifier Using Forward Combining Technique," *IEEE Microw. Wirel. Compon. Lett.*, vol. 20, no. 12, pp. 672-674, Dec. 2010.

[64]   Y. -F. Tsao and H. -T. Hsu, "A 52–58 GHz Power Amplifier With 18.6-dBm Saturated Output Power for Space Applications," *IEEE Trans. Circuits Syst. II: Exp. Briefs*, vol. 68, no. 6, pp. 1927-1931, June 2021.

[65]   Richard Lossy, Hervé Blanck and Joachim Würfl, "Reliability studies on GaN HEMTs with sputtered Iridium gate module", Microelectronics Reliability, vol. 52, no. 9-10, pp. 2144 - 2148, Sep - Oct. 2012.

[66]   Keysight   Technologies: 'Noise   Source   Operating   and   Service   Manual', https://www.keysight.com/main/gated.jspx?lb=1&gatedId=1000002290-1:epsg:man&cc=TW&lc=cht&parentContId=375206&parentContType=pt&parentNid=-536902744.536883696&fileType=VIEWABLE

[67]   M. Rudolph, N. Chaturvedi, K. Hirche, J. Wurfl, W. Heinrich and G. Trankle, "Highly Rugged 30 GHz GaN Low-Noise Amplifiers," *IEEE Microw. Wirel. Compon. Lett.*, vol. 19, no. 4, pp. 251-253, April 2009.

[68]   X. Tong *et al.*, "A 22–30-GHz GaN Low-Noise Amplifier With 0.4–1.1-dB Noise Figure," *IEEE Microw. Wirel. Compon. Lett.*, vol. 29, no. 2, pp. 134-136, Feb. 2019.

[69]   X. Tong *et al.*, "An 18–56-GHz Wideband GaN Low-Noise Amplifier With 2.2–4.4-dB Noise Figure," *IEEE Microw. Wirel. Compon. Lett.*, vol. 30, no. 12, pp. 1153-1156, Dec. 2020.

[70]   H. P. Moyer *et al.*, "Q-Band GaN MMIC LNA Using a 0.15µm T-Gate Process," in *Proc. 2008 IEEE Compound Semi. Integr. Circuits Symp.*, 2008, pp. 1-4.

[71]   M. Sato *et al.*, "Q-Band InAlGaN/GaN LNA using current reuse topology," in *Proc. 2016 IEEE MTT-S Int. Microw. Symp. (IMS)*, 2016, pp. 1-4.

[72]   Y. Chen, Y. Lin, J. Lin and H. Wang, "A Ka-Band Transformer-Based Doherty Power Amplifier for Multi-Gb/s Application in 90-nm CMOS," *IEEE Microw. Wirel. Compon. Lett.*, vol. 28, no. 12, pp. 1134-1136, Dec. 2018.

[73]   S. Shakib, H. Park, J. Dunworth, V. Aparin and K. Entesari, "A Highly Efficient and Linear Power Amplifier for 28-GHz 5G Phased Array Radios in 28-nm CMOS," *IEEE J. Solid-State Circuits*, vol. 51, no. 12, pp. 3020-3036, Dec. 2016.

[74]   V. Qunaj and P. Reynaert, "A Compact Ka-Band Transformer-Coupled Power Amplifier for 5G in 0.15um GaAs," in *Proc. 2019 IEEE BiCMOS Compound Semiconductor Integr. Circuits Techn. Symp. (BCICTS)*, 2019, pp. 1-4.

[75]   H. Wang, C. Rosa and K. Pedersen, "Performance Analysis of Downlink Inter-Band Carrier Aggregation in LTE-Advanced," in *Proc. 2011 IEEE Vehicular Techn. Conf. (VTC Fall)*, 2011, pp. 1-5.

[76]   C. S. Park, L. Sundström, A.Wallén, A. Khayrallah, "Carrier aggregation for LTE-advanced: Design challenges of terminals," *IEEE Commun. Mag.*, vol. 51, no. 12, pp. 76-84, Dec. 2013.

[77]   Nokia and Optus, "Press Release: Nokia and Optus demonstrate 10 Gbps from a live 5G site," Apr. 7, 2021. [Online] Available: https://www.nokia.com/about-us/news/releases/2021/04/06/nokia-and-optus-demonstrate-10-gbps-from-a-live-5g-site/

[78]   T. Cappello, A. Duh, T. W. Barton and Z. Popovic, "A Dual-Band Dual-Output Power Amplifier for Carrier Aggregation," *IEEE Trans. Microw. Theory Techn.*, vol. 67, no. 7, pp. 3134-3146, July 2019.

[79]   G. Lv, W. Chen, X. Chen and Z. Feng, "An Energy-Efficient Ka/Q Dual-Band Power Amplifier MMIC in 0.1-µm GaAs Process," *IEEE Microw. Wirel. Compon. Lett.*, vol. 28, no. 6, pp. 530-532, June 2018.

[80]   C. Huynh and C. Nguyen, "New Technique for Synthesizing Concurrent Dual-Band Impedance-Matching Filtering Networks and 0.18-µm SiGe BiCMOS 25.5/37-GHz Concurrent Dual-Band Power Amplifier," *IEEE Trans. Microw. Theory Techn.*, vol. 61, no. 11, pp. 3927-3939, Nov. 2013.

[81]   S. Hu, F. Wang and H. Wang, "A 28-/37-/39-GHz Linear Doherty Power Amplifier in Silicon for 5G Applications," *IEEE J. Solid-State Circuits*, vol. 54, no. 6, pp. 1586-1599, June 2019.

[82]   H. Xie *et al.*, "A High-Efficiency 28/39 GHz Dual-Band Power Amplifier MMIC for 5G Communication," *IEEE Microw. Wirel. Compon. Lett.*, Early Access, doi: 10.1109/LMWC.2021.3095659.

[83]   B.-W. Huang, Z.-H. Fu and K.-Y. Lin, "A 28/39 GHz Dual-Band Power Amplifier Using Optimal Matching Contour in GaAs pHEMT," in *Proc. 2021 IEEE Int. Symp. Radio-Frequency Integr. Techn. (RFIT)*, 2021, pp. 1-3.

[84]   3GPP Base Station (BS) Radio Transmission and Reception, "Technical Specification (TS) 38.104, 3rd Generation Partnership Project (3GPP)," Version 17.2.0, July 2021. [Online] Available: https://www.3gpp.org/ftp/Specs/archive/38_series/38.104/38104-h20.zip

[85]   C. Sun, H. Zheng, L. Zhang and Y. Liu, "Analysis and Design of a Novel Coupled Shorting Strip for Compact Patch Antenna With Bandwidth Enhancement," *IEEE Antennas Wirel. Propag. Lett.*, vol. 13, pp. 1477-1481, 2014.

[86]   CST Microwave Studio. Accessed: Aug. 1, 2017. [Online]. Available: http://www.cst.com/products/cstmws

[87]   C. Fan, B. Wu, Y. Hu, Y. Zhao and T. Su, "Millimeter-Wave Pattern Reconfigurable Vivaldi Antenna Using Tunable Resistor Based on Graphene," *IEEE Trans. Antennas Propag.*, vol. 68, no. 6, pp. 4939-4943, June 2020.

[88]   M. Rastegari, A. Olfat and E. Amini, "Analogue beamforming solution for beam-alignment problem," *IET Comm.*, vol. 14, no. 15, pp. 2576-2583, 15 9 2020.

[89]   P. K. Bailleul, "A New Era in Elemental Digital Beamforming for Spaceborne Communications Phased Arrays," *Proc. IEEE*, vol. 104, no. 3, pp. 623-632, March 2016.

[90]   S. Han, C. I, Z. Xu and S. Wang, "Reference Signals Design for Hybrid Analog and Digital Beamforming," *IEEE Comm. Lett.*, vol. 18, no. 7, pp. 1191-1193, July 2014.

[91]   Yu-Hsuan Lin and H. Wang, "A low phase and gain error passive phase shifter in 90 nm CMOS for 60 GHz phase array system application," in *Proc. 2016 IEEE MTT-S Int. Microw. Symp. (IMS)*, 2016, pp. 1-4.

[92]   C. Chen, Y. Wang, Y. Lin, Y. Hsiao, Y. Wu and H. Wang, "A 36–40 GHz full 360° ultra-low phase error passive phase shifter with a novel phase compensation technique," in *Proc. 2017 47th Euro. Microw. Conf. (EuMC)*, 2017, pp. 1245-1248.

[93]   X. Quan *et al.*, "A 52–57 GHz 6-Bit Phase Shifter With Hybrid of Passive and Active Structures," *IEEE Microw. Wirel. Compon. Lett.*, vol. 28, no. 3, pp. 236-238, March 2018.

[94]   Woosung Lee, Jaeheung Kim, Choon Sik Cho, and Young Joong Yoon, "Beamforming Lens Antenna on a High Resistivity Silicon Wafer for 60 GHz WPAN," *IEEE Trans. Antennas Propag.*, vol. 58, no. 3, Mar. 2010.

[95]   H. Chu, Y. X. Guo and Z. Wang, "60 GHz LTCC Wideband Vertical Off Center Dipole Antenna and Arrays," *IEEE Trans. Antennas Propag.*, vol. 61, no. 1, pp. 153 161, Jan. 2013.

[96]   S. Chao, H. Wang, C. Su and J. G. J. Chern, "A 50 to 94-GHz CMOS SPDT Switch Using Traveling-Wave Concept," *IEEE Microw. Wirel. Compon. Lett.*, vol. 17, no. 2, pp. 130-132, Feb. 2007.

[97]   H. Chang and C. Chan, "A Low Loss High Isolation DC-60 GHz SPDT Traveling-Wave Switch With a Body Bias Technique in 90 nm CMOS Process," *IEEE Microw. Wirel. Compon. Lett.*, vol. 20, no. 2, pp. 82-84, Feb. 2010.

[98]   K. T. Trinh, H. Kao, H. Chiu and N. C. Karmakar, "A Ka-Band GaAs MMIC Traveling-Wave Switch With Absorptive Characteristic," *IEEE Microw. Wirel. Compon. Lett.*, vol. 29, no. 6, pp. 394-396, June 2019.

[99]   D. Jahn, R. Reuter, Y. Yin and J. Feige, "Characterization and Modeling of Wire Bond Interconnects up to 100 GHz," in *Proc. 2006 IEEE Compound Semiconductor Integr. Circuit Symp.*, 2006, pp. 111-114.

Innovationen mit Mikrowellen und Licht
Forschungsberichte aus dem Ferdinand-Braun-Institut, Leibniz-Institut für Höchstfrequenztechnik

Herausgeber: Prof. Dr. G. Tränkle

Band 1: **Thorsten Tischler**
Die Perfectly-Matched-Layer-Randbedingung in der
Finite-Differenzen-Methode im Frequenzbereich:
Implementierung und Einsatzbereiche
ISBN: 3-86537-113-2, 19,00 EUR, 144 Seiten

Band 2: **Friedrich Lenk**
Monolithische GaAs FET- und HBT-Oszillatoren
mit verbesserter Transistormodellierung
ISBN: 3-86537-107-8, 19,00 EUR, 140 Seiten

Band 3: **R. Doerner, M. Rudolph (eds.)**
Selected Topics on Microwave Measurements,
Noise in Devices and Circuits, and Transistor Modeling
ISBN: 3-86537-328-3, 19,00 EUR, 130 Seiten

Band 4: **Matthias Schott**
Methoden zur Phasenrauschverbesserung von
monolithischen Millimeterwellen-Oszillatoren
ISBN: 978-3-86727-774-0, 19,00 EUR, 134 Seiten

Band 5: **Katrin Paschke**
Hochleistungsdiodenlaser hoher spektraler Strahldichte
mit geneigtem Bragg-Gitter als Modenfilter (α-DFB-Laser)
ISBN: 978-3-86727-775-7, 19,00 EUR, 128 Seiten

Band 6: **Andre Maaßdorf**
Entwicklung von GaAs-basierten Heterostruktur-Bipolartransistoren
(HBTs) für Mikrowellenleistungszellen
ISBN: 978-3-86727-743-3, 23,00 EUR, 154 Seiten

Band 7: **Prodyut Kumar Talukder**
Finite-Difference-Frequency-Domain Simulation of Electrically
Large Microwave Structures using PML and Internal Ports
ISBN: 978-3-86955-067-1, 19,00 EUR, 138 Seiten

Band 8: **Ibrahim Khalil**
Intermodulation Distortion in GaN HEMT
ISBN: 978-3-86955-188-3, 23,00 EUR, 158 Seiten

Band 9: **Martin Maiwald**
Halbleiterlaser basierte Mikrosystemlichtquellen für die Raman-Spektroskopie
ISBN: 978-3-86955-184-5, 19,00 EUR, 134 Seiten

Band 10: **Jens Flucke**
Mikrowellen-Schaltverstärker in GaN- und GaAs-Technologie
Designgrundlagen und Komponenten
ISBN: 978-3-86955-304-7, 21,00 EUR, 122 Seiten

Cuvillier Verlag
Internationaler wissenschaftlicher Fachverlag

Innovationen mit Mikrowellen und Licht
Forschungsberichte aus dem Ferdinand-Braun-Institut, Leibniz-Institut für Höchstfrequenztechnik

Herausgeber: Prof. Dr. G. Tränkle

Band 11: Harald Klockenhoff
Optimiertes Design von Mikrowellen-Leistungstransistoren
und Verstärkern im X-Band
ISBN: 978-3-86955-391-7, 26,75 EUR, 130 Seiten

Band 12: Reza Pazirandeh
Monolithische GaAs FET- und HBT-Oszillatoren
mit verbesserter Transistormodellierung
ISBN: 978-3-86955-107-8, 19,00 EUR, 140 Seiten

Band 13: Tomas Krämer
High-Speed InP Heterojunction Bipolar Transistors
and Integrated Circuits in Transferred Substrate Technology
ISBN: 978-3-86955-393-1, 21,70 EUR, 140 Seiten

Band 14: Phuong Thanh Nguyen
Investigation of spectral characteristics of solitary
diode lasers with integrated grating resonator
ISBN: 978-3-86955-651-2, 24,00 EUR, 156 Seiten

Band 15: Sina Riecke
Flexible Generation of Picosecond Laser Pulses in the Infrared and
Green Spectral Range by Gain-Switching of Semiconductor Lasers
ISBN: 978-3-86955-652-9, 22,60 EUR, 136 Seiten

Band 16: Christian Hennig
Hydrid-Gasphasenepitaxie von versetzungsarmen und freistehenden
GaN-Schichten
ISBN: 978-3-86955-822-6, 27,00 EUR, 162 Seiten

Band 17: Tim Wernicke
Wachstum von nicht- und semipolaren InAlGaN-Heterostrukturen
für hocheffiziente Licht-Emitter
ISBN: 978-3-86955-881-3, 23,40 EUR, 138 Seiten

Band 18: Andreas Wentzel
Klasse-S Mikrowellen-Leistungsverstärker mit GaN-Transistoren
ISBN: 978-3-86955-897-4, 29,65 EUR, 172 Seiten

Band 19: Veit Hoffmann
MOVPE growth and characterization of (In,Ga)N quantum structures
for laser diodes emitting at 440 nm
ISBN: 978-3-86955-989-6, 18,00 EUR, 118 Seiten

Band 20: Ahmad Ibrahim Bawamia
Improvement of the beam quality of high-power broad area
semiconductor diode lasers by means of an external resonator
ISBN: 978-3-95404-065-0, 21,00 EUR, 126 Seiten

Cuvillier Verlag
Internationaler wissenschaftlicher Fachverlag

Innovationen mit Mikrowellen und Licht
Forschungsberichte aus dem Ferdinand-Braun-Institut, Leibniz-Institut für Höchstfrequenztechnik

Herausgeber: Prof. Dr. G. Tränkle

Band 21:
Agnietzka Pietrzak
Realization of High Power Diode Lasers with Extremely Narrow Vertical Divergence
ISBN: 978-3-95404-066-7, 27,40 EUR, 144 Seiten

Band 22:
Eldad Bahat-Treidel
GaN-based HEMTs for High Voltage Operation
Design, Technology and Characterization
ISBN: 978-3-95404-094-0, 41,10 EUR, 220 Seiten

Band 23:
Ponky Ivo
AlGaN/GaN HEMTs Reliability:
Degradation Modes and Anslysis
ISBN: 978-3-95404-259-3, 23,55 EUR, 132 Seiten

Band 24:
Stefan Spießberger
Compact Semiconductor-Based Laser Sources
with Narrow Linewidth and High Output Power
ISBN: 978-3-95404-261-6, 24,15 EUR, 140 Seiten

Band 25:
Silvio Kühn
Mikrowellenoszillatoren für die Erzeugung von atmosphärischen Mikroplasmen
ISBN: 978-3-95404-378-1, 21,85 EUR, 112 Seiten

Band 26:
Sven Schwertfeger
Experimentelle Untersuchung der Modensynchronisation in Multisegment-Laserdioden zur Erzeugung kurzer optischer Pulse bei einer Wellenlänge von 920 nm
ISBN: 978-3-95404-471-9, 29,45 EUR, 150 Seiten

Band 27:
Christoph Matthias Schultz
Analysis and mitigation of the factors limiting the effiency of high power distributed feedback diode lasers
ISBN: 978-3-95404-521-1, 68,40 EUR, 388 Seiten

Band 28:
Luca Redaelli
Design and fabrication of GaN-based laser diodes for single-mode and narrow-linewidth applications
ISBN: 978-3-95404-586-0, 29,70 EUR, 176 Seiten

Band 29:
Martin Spreemann
Resonatorkonzepte für Hochleistungs-Diodenlaser mit ausgedehnten lateralen Dimensionen
ISBN: 978-3-95404-628-7, 25,15 EUR, 128 Seiten

Cuvillier Verlag
Internationaler wissenschaftlicher Fachverlag

Innovationen mit Mikrowellen und Licht
Forschungsberichte aus dem Ferdinand-Braun-Institut,
Leibniz-Institut für Höchstfrequenztechnik

Herausgeber: Prof. Dr. G. Tränkle

Band 30: **Christian Fiebig**
Diodenlaser mit Trapezstruktur und hoher Brillanz für die Realisierung
einer Frequenzkonversion auf einer mikro-optischen Bank
ISBN: 978-3-95404-690-4, 26,30 EUR, 140 Seiten

Band 31: **Viola Küller**
Versetzungsreduzierte AIN- und AlGaN-Schichten
als Basis für UV LEDs
ISBN: 978-3-95404-741-3, 34,40 EUR, 164 Seiten

Band 32: **Daniel Jedrzejczyk**
Efficient frequency doubling of near-infrared diode lasers
using quasi phase-matched waveguides
ISBN: 978-3-95404-958-5, 27,90 EUR, 134 Seiten

Band 33: **Sylvia Hagedorn**
Hybrid-Gasphasenepitaxie zur Herstellung von Aluminiumgalliumnitrid
ISBN: 978-3-95404-985-1, 38,00 EUR, 176 Seiten

Band 34: **Alexander Kravets**
Advanced Silicon MMICs for mm-Wave Automotive Radar Front-Ends
ISBN: 978-3-95404-986-8, 31,90 EUR, 156 Seiten

Band 35: **David Feise**
Longitudinale Modenfilter für Kantanemitter im roten Spektralbereich
ISBN: 978-3-7369-9116-3, 39,20 EUR, 168 Seiten

Band 36: **Ksenia Nosaeva**
Indium phosphide HBT in thermally optimized periphery for
applications up to 300GHZ
ISBN: 978-3-7369-287-0, 42,00 EUR, 154 Seiten

Band 37: **Muhammad Maruf Hossain**
Signal Generation for Millimeter Wave and THZ Applications
in InP-DHBT and InP-on-BiCMOS Technologies
ISBN: 978-3-7369-9335-8, 35,60 EUR, 136 Seiten

Band 38: **Sirinpa Monayakul**
Development of Sub-mm Wave Flip-Chip Interconnect
ISBN: 978-3-7369-9410-2, 44,00 EUR, 146 Seiten

Band 39: **Moritz Brendel**
Charakterisierung und Optimierung von (Al, Ga) N-basierten
UV-Photodetektoren
ISBN: 978-3-7369-9465-2, 49,90 EUR, 196 Seiten

Band 40: **Erdenetsetseg Luvsandamdin**
Development of micro-integrated diode lasers for precision quantum optics
experiments in space
ISBN: 978-3-7369-9479-9, 39,00 EUR, 126 Seiten

Cuvillier Verlag
Internationaler wissenschaftlicher Fachverlag

Innovationen mit Mikrowellen und Licht
Forschungsberichte aus dem Ferdinand-Braun-Institut, Leibniz-Institut für Höchstfrequenztechnik

Herausgeber: Prof. Dr. G. Tränkle

Band 41:
Thi Nghiem Vu
Development and analysis of diode laser ns-MOPA systems
for high peak power application
ISBN: 978-3-7369-9480-5, 38,80 EUR, 138 Seiten

Band 42:
Christian Bansleben
Differentieller Mikrowellen-Leistungsoszillator für die Realisierung
ultrakompakter Plasmaquellen in Matrixanordnung
ISBN: 978-3-7369-9530-7, 34,90 EUR, 136 Seiten

Band 43:
Martin Winterfeldt
Investigation of slow-axis beam quality degradation
in high-power broad area diode lasers
ISBN: 978-3-7369-9733-2, 39,90 EUR, 158 Seiten

Band 44:
Jonathan Decker
Investigation of monolithically integrated spectral stabilization
in high-brightness broad area diode lasers
ISBN: 978-3-7369-9798-1, 49,50 EUR, 174 Seiten

Band 45:
Andreea Cristina Andrei
Untersuchung und Optimierung robuster und hochlinearer
rauscharmer Verstärker in GaN-Technologie
ISBN: 978-3-7369-9810-0, 39,90 EUR, 148 Seiten

Band 46:
Peng Luo
GaN HEMT Modeling Including Trapping Effects Based on
Chalmers Model and Pulsed S-Parameter Measurements
ISBN: 978-3-7369-9906-0, 48,00 EUR, 160 Seiten

Band 47:
Jörg Jeschke
Entwicklung von optisch pumpbaren UVC-Lasern auf AlGaN-Basis
Chalmers Model and Pulsed S-Parameter Measurements
ISBN: 978-3-7369-9918-3, 44,90 EUR, 176 Seiten

Band 48:
Nikolai Wolff
Wideband GaN Microwave Power Amplifiers with Class-G
Supply Modulation
ISBN: 978-3-7369-9931-2, 44,90 EUR, 170 Seiten

Band 49:
Carlo Frevert
Optimization of broad-area GaAs diode lasers for high powers
and high efficiences in the temperature range 200-220 K
ISBN: 978-3-7369-9944-2, 44,90 EUR, 174 Seiten

Band 50:
Bassem Arar
GaAs-based components for photonic integrated circuits
ISBN: 978-3-7369-9976-3, 43,60 EUR, 152 Seiten

Cuvillier Verlag
Internationaler wissenschaftlicher Fachverlag

Innovationen mit Mikrowellen und Licht
Forschungsberichte aus dem Ferdinand-Braun-Institut, Leibniz-Institut für Höchstfrequenztechnik

Herausgeber: Prof. Dr. G. Tränkle

Cuvillier Verlag
Internationaler wissenschaftlicher Fachverlag

Innovationen mit Mikrowellen und Licht
Forschungsberichte aus dem Ferdinand-Braun-Institut,
Leibniz-Institut für Höchstfrequenztechnik

Herausgeber: Prof. Dr. G. Tränkle

Band 61:
Max Schiemangk
Ein Lasersystem für Experimente mit Quantengasen unter
Schwerelosigkeit
ISBN: 978-3-7369-7293-3, 49,90 EUR, 166 Seiten

Band 62:
Thorben Kaul
Epitaxial Design Optimizations for Increased Efficiency in
GaAs-Based High Power Diode Lasers
ISBN: 978-3-7369-7396-1, 38,90 EUR, 136 Seiten

Band 63:
Pietro della Casa
Two Step MOPVE, in-situ etching and buried implantations:
applications to the realization of GaAs laser diodes
ISBN: 978-3-7369-7397-8, 72,90 EUR, 250 Seiten

Band 64:
Heike Christopher
A compact mode-locked diode laser system for high prescision frequency
comparison experiments
ISBN: 978-3-7369-7399-2, 59,90 EUR, 206 Seiten

Band 65:
Rimma Zhytnytska
Design von GaN Transistoren für leistungselektronische Anwendungen
ISBN: 978-3-7369-7413-5, 57,90 EUR, 182 Seiten

Band 66:
Sebastian Walde
AlN base layers for UV LEDs
ISBN: 978-3-7369-7451-7, 44,90 EUR, 156 Seiten

Band 67:
Jan Ruschel
Ursachen der stromgetriebenen Degradation von UV-LEDs
ISBN: 978-3-7369-7542-2, 49,90 EUR, 180 Seiten

Band 68:
Erik Freier
Optimierung der Prozesstechnologie und Steigerung der Zuverlässungkeit
und Lebensdauer von (InAlGa)N-basierten Halbleiterlaserdioden
ISBN: 978-3-7369-7604-7, 49,90 EUR, 178 Seiten

Band 69:
Norman Ruhnke
A deep ultraviolet laser light source by frequency doubling
of GaN based external cavity diode laser radiation
ISBN: 978-3-7369-7613-9, 39,90 EUR, 144 Seiten

Band 70:
Matthias M. Karow
Broad-Area Laser Bars for 1 kW-Emission.
Demonstrating increased efficiency and narrow far field
ISBN: 978-3-7369-7626-9, 39,50 EUR, 142 Seiten

Cuvillier Verlag
Internationaler wissenschaftlicher Fachverlag